HEAD

HEADS UP!

Early Warning Systems for Climate-, Water- and Weather-Related Hazards

Edited by Michael H. Glantz

United Nations
University Press

TOKYO • NEW YORK • PARIS

United Nations University Press
United Nations University, 53-70, Jingumae 5-chome,
Shibuya-ku, Tokyo 150-8925, Japan
Tel: +81-3-5467-1212 Fax: +81-3-3406-7345
E-mail: sales@hq.unu.edu general enquiries: press@hq.unu.edu
http://www.unu.edu

United Nations University Office at the United Nations, New York
2 United Nations Plaza, Room DC2-2062, New York, NY 10017, USA
Tel: +1-212-963-6387 Fax: +1-212-371-9454
E-mail: unuona@ony.unu.edu

United Nations University Press is the publishing division of the United Nations University.

Cover design by Mori Design Incl. Tokyo

Printed in Hong Kong

ISBN 978-92-808-1169-8

Library of Congress Cataloging-in-Publication Data

Heads up! : early warning systems for climate, water and weather-related hazards / edited by Michael H. Glantz.
p. cm.
Includes bibliographical references and index.
ISBN 978-9280811698 (pbk.)
1. Climatic changes—Forecasting. 2. Climatic changes—Forecasting—Environmental aspects. 3. Geographic information systems. 4. Weather forecasting I. Glantz, Michael H.
QC981.8.C5H43 2009
363.34—dc22 2009016299

Contents

Tables and figures

Tables

Figures

Contributors

Jennifer Boehnert – GIS Coordinator, NCAR
Thomas Bogdan – Director, Space Weather Prediction Center, Space Environment Center (NOAA National Weather Service), Boulder, Colorado
Bureau of Meteorology, Australia – Melbourne, Australia
Stanley Changnon – Emeritus Climatologist, Illinois State Water Survey
S. H. M. Fakhruddin – Early Warning Systems Technical Specialist, ADPC, Bangkok
Michael H. Glantz – Director, Consortium for Capacity Building, University of Colorado
Eve Gruntfest – Director, Social Science Woven into Meteorology, University of Oklahoma
Mary Hayden – Postdoctoral Fellow, ISSE, NCAR
Stefanie Herrmann – Postdoctoral Fellow, ASP, NCAR
Ilan Kelman – Scientist, Center for International Climate and Environmental Research (CICERO), Oslo, Norway
Joanie Kleypas – Scientist, NCAR
Janet Larsen – Director of Research, Earth Policy Institute, Washington, DC.
Margaret LeMone – Scientist, NCAR
Michael McPhaden – Scientist, PMEL, NOAA, Seattle, Washington
Paulette Middleton – Scientist, Panorama Pathways, Boulder, Colorado
Rebecca Morss – Scientist, NCAR
Susanne Moser – Director, Susanne Moser Research and Consulting, Santa Cruz, California
Lino Naranjo Diaz – MeteoGalicia, A Coruña, Spain
Natural Resources Canada – Ottawa and Ontario, Canada
Douglas Pattie – Scientist, International Strategy for Disaster Reduction, Bonn, Germany

Roger Pielke, Jr. – Professor, Center for Science and Technology Policy Research (CSTPR), University of Colorado
Gregory Pierce – Research Associate, Consortium for Capacity Building, University of Colorado
Kevin Trenberth – Scientist, NCAR
UN Food and Agriculture Organization – Rome, Italy
Tsegay Wolde-Georgis – Consortium for Capacity Building, University of Colorado
Qian Ye – Consortium for Capacity Building and Beijing Normal University, China

Foreword

An early warning system is an important tool in a government's program to reduce risk to hazards and an essential requirement to achieve sustainable development. In fact sustainable development prospects are very dependent on the effectiveness of early warning systems. In a broad sense, these systems are a social process where a heavy responsibility is placed on national institutions to provide resources and enhance local community ownership and participation through the local institutions to prevent, mitigate and respond to disasters. However, if the capacity to issue or follow up on early warnings and disaster risk reduction plans is not adequate, then early warning systems will not support effectively community and individuals' needs.

To ensure that early warning leads to early action, warning information must be viewed as reliable and credible by all parties. Busy decision-makers swamped with information tend to gravitate towards reliance on simple, straightforward messages. There is always political pressure for timely use of information because of the inevitable conflicts of interests over allocation of relief resources.

Disaster risk reduction begins with information. The more we understand natural hazards and related technological and environmental disasters as well as their causes and consequences on societies, the more we are able to better prepare to reduce their risks. The more that decision-makers at all levels of society commit themselves to disaster reduction policies and actions, the sooner vulnerable communities will be safer and more resilient.

Preparedness and prevention are inextricably bound up with the use or misuse of information. A review of various types of early warning systems offers insight into other, less clear-cut cases of information used to trigger public action. In many respects, early warning systems are reflective of how information and knowledge are – or are not – used in development planning. Most early warning and response systems are highly centralized. Information has to be aggregated to fit within a bureaucratic structure, often losing the knowledge about local people's coping strategies. The decisions to respond and to mobilize resources are often taking place hundreds, even thousands of miles from where help is needed, by people who are far removed from what is happening on the ground and with little sense of urgency or understanding of local conditions.

There are different kinds of early warning systems, designed to fulfill different functions, and to fit different natural hazards, such as floods, landslides, food security, famine drought, avian flu, tsunamis, cyclones, volcanoes and wildland fires, among other. One of the major challenges in developing a global approach to these hazards is how best to integrate systems operating at different locations and at different scales in order to provide a more coordinated, comprehensive, usable and effective warning.

But, however large or complex the formal early warning system is, there is an even larger early warning network which encompasses many more elements of society than one might realize. In this new UNU publication, Michael Glantz has put together a straightforward, usable collection of scenarios, lessons and good practices as well as examples of early warning systems that have been translated into practical experiences at the national level. Clearly, we need to invest more to enhance national capacity in public awareness, education and information. Early warning of hazards combined with the early warnings of underlying societal problems and processes can lead to a strengthening of resilience and a reduction in vulnerability. As this work points out, an effective early warning system is a necessary investment in disaster risk reduction.

Salvano Briceno

Director

United Nations International Strategy for Disaster Reduction

Acknowledgements

The production of this volume was truly a team effort. First of all, I sincerely appreciate the several prototypes designed by Anne Blondeel-Oman. D. Jan Stewart, my assistant for 29 years, was also instrumental in the preparation of this publication. Most recently, Gregory Pierce was the final copyeditor and research associate *par excellence* in the preparation and layout of countless drafts of the text and images. I am very grateful for the constructive comments of the anonymous reviewer.

A final acknowledgement must be made of the sincere efforts in just about every country of hundreds of organizations involved in formal and informal early warning activities. Their overriding missions are to protect at-risk populations from harm, especially harm that might result from climate-, water- and weather-related hazards.

Michael H. Glantz
Director, Consortium for Capacity Building
Institute for Arctic and Alpine Research (INSTAAR)
University of Colorado
Boulder, Colorado (USA)

March 2009

Abbreviations

AI	avian influenza
AQI	Air Quality Index
ATWC	West Coast/Alaska Tsunami Warning Center
BUI	Build up Index
CHAMP	NOAA Coral Health and Monitoring Program
CMI	Crop Moisture Index
CPC	US Climate Prediction Center
CREWS	Coral Reef Early Warning System
DC	Drought Code
DHF	dengue haemorrhagic fever
DHW	Degree Heating Week
DMC	Duff Moisture Code
DSS	dust and sandstorm
DVSI	Denver Visibility Standard Index
EEA	European Environment Agency
ENSO	El Niño Southern Oscillation
EPA	US Environmental Protection Agency
ESCAP	Economic and Social Commission for Asia and the Pacific
Eumenet	Network of European Meteorological Services
EWS	early warning system
FAA	US Federal Aviation Administration
FAO	UN Food and Agriculture Organization
FAO/GIEWS	FAO Global Information and Early Warning System
FEMA	US Federal Emergency Management Agency
FEWS-Net	Famine Early Warning System Network
FFMC	Fine Fuel Moisture Code

FWI	Forest Fire Weather Index
GAINS	Global Avian Influenza Network for Surveillance
GHG	greenhouse gas
GIS	geographic Information Systems
GLEWS	Global Early Warning System
GPS	Global Positioning System
HELI	WHO/UNEP Health and Environment Linkages Initiative
HIA	Health Impact Assessment
HPAI	highly pathogenic avian influenza
ICG/IOTWS	Intergovernmental Coordination Group for the Indian Ocean Tsunami Warning and Mitigation System
ICON	NOAA Integrated Coral Observing Network
IOC	Intergovernmental Oceanographic Commission
IPCC	Intergovernmental Panel on Climate Change
ISDR	International Strategy for Disaster Reduction
ISI	Initial Spread Index
IVM	Integrated Vector Management
MARA/ARMA	Mapping Malaria Risk in Africa initiative
NAMS	National Agricultural Monitoring System (Australia)
NASA	US National Aeronautics and Space Administration
NGO	non-governmental organization
NMA	National Meteorological Agency (Ethiopia)
NOAA	US National Oceanic and Atmospheric Administration
NWS	US National Weather Service
MoH	Ministry of Health (Ethiopia)
MSD	Mesoscale Discussion
OIE	World Organization for Animal Health
PDA	personal digital assistant
PDSI	Palmer Drought Severity Index

ppmv	parts per million by volume
PTWC	Pacific Tsunami Warning Center
RANET	Radio and Internet for the Communication of Hydro-Meteorological Information for Rural Development
RSMC	Regional Specialized Meteorological Centre
RVF	Rift Valley fever
SEC	NOAA Space Environment Center
SFDH	State Flood Control and Drought Relief Headquarters (China)
SPC	US NWS Storm Prediction Center
SPF	Sun Protection Factor
SSA	sub-Saharan Africa
SST	sea surface temperature
SWOC	Strengths, Weaknesses, Opportunities and Constraints
UNDP	UN Development Programme
UNEP	UN Environment Programme
UNESCO	UN Economic, Social and Cultural Organization
UNFAO	UN Food and Agriculture Organization
USAID	US Agency for International Development
USGS	US Geological Survey
UV	ultraviolet
VSI	Visibility Standard Index
WFP	World Food Programme
WHO	World Health Organization
WMO	World Meteorological Organization

1

Introduction

Interest in early warning

Concerned about impending or likely threats and the problems they may face as a result, all governments, corporations, groups and individuals are interested in early warning in one form or another. At least in theory, the more advanced the warning they are given, the better off they are because they then have ample time to prepare for – at minimum – and hopefully respond effectively to the potential impacts of natural or human-induced threats. As the adage goes in many, if not all, cultures: "To be forewarned is to be forearmed." Yet, as history continues to show, warnings alone are not enough; they must be coupled with a government's willingness and ability to respond.

Many early warning systems (EWSs) exist today to inform the general public, governments and businesses such as insurance companies and grain producers about impending climate-, water- and weather-related hazards, among other natural and human-made threats. The experiences and insights identified around the globe through the use of these EWSs have helped to forearm officials and other decision-makers in various governmental and non-governmental organizations (NGOs) about how to prepare for and communicate effective warnings about future threats. Aside from being the right thing to do – that is, to protect citizens from harm – it is a moral necessity for governments and NGOs to anticipate and respond to threats by protecting their at-risk populations and ecosystems. Sharing experiences also helps to educate the media and the general public about how to interpret warnings and apply them to their own local needs. In other words, possessing information is a source of power, and sharing information empowers

Heads Up! Early Warning Systems for Climate-, Water- and Weather-Related Hazards, Michael H. Glantz (ed.), United Nations University Press, 2009, ISBN 978-92-808-1169-8

people by helping them to understand the value, use and limitations of warnings for foreseeable hazards.

Many organizations, from the local to the global, are responsible for early warnings of impending threats to life, limb and property, and the UN system has been given special charge by governments worldwide to foster awareness and progress in the development and effective use of EWSs. Organizations like the WMO, WHO, UNEP, UNDP, UNFAO, UNESCO, IOC, ISDR, WFP, FAO/GIEWS and FEWS-Net, among others, are deeply involved in the various aspects of early warning relevant to their institutional jurisdictions Humanitarian NGOs are also dependent on EWS output for their effective operations. It is important to keep in mind that there is likely a mix of several formal and informal warning systems focused on similar climate-, water- and weather-related hazards operating at the same time in any given area. Collectively, these systems provide a first defence against a variety of hazards to those in harm's way.

The "Precautionary Principle" as the basis of early warning

In the 1980s the notion of the "Precautionary Principle" began to take hold in international environmental discussions. Based on the view that the scientific uncertainties surrounding an environmental stress should not be used as an excuse for inaction in addressing that stress, the principle supports consideration of the old maxim, "better to be safe now than sorry later", especially if an environmental stress could become critical and have irreversible and costly consequences in time or space. The European Environment Agency, for example, published a collection of studies documenting cases in which the Precautionary Principle was not applied. The publication, entitled *Late Lessons from Early Warnings* (EEA, 2002), gathers information on hazards raised by human economic activities and tracks the use of that information in actions taken to give better protection to the environment and the health of species and ecosystems dependent upon it. In doing so, the EEA (ibid.) "aims to contribute to better and more accessible science-based information and more effective stakeholder participation in the governance of economic activity to help minimize future environmental and health costs and maximize innovation".

Human influence on global climate appears destined, however, to be only the latest global environmental stress (some say "insult") to become, regrettably, just another case of early warnings accompanied by late lessons, though time still remains to prevent, mitigate or adapt to some of climate change's worst causes and impacts. Mindful application of the Precautionary Principle can help avert this swiftly approaching outcome.

What constitutes an early warning system (EWS)?

An early warning system (EWS) is made up of several components and is not well represented as being only the formulation and issuance of a warning. A holistic EWS includes the formulation of the warning, the issuance of the warning, the reception of and response to the warning, and finally feedback to those who developed and issued the warning in the first place. Each component has to be considered in evaluating the system. A weakness in any part of this chain of steps, from warning preparations to responses, can render an early warning system ineffective – an outcome critical to avoid because a system that does not warn will not be taken seriously. Furthermore, an effective EWS must always contain a well-functioning feedback loop, so those responsible for developing and issuing warnings can determine the value of specific types of warnings to at-risk populations and also evaluate the effectiveness of their systems in general (a hindcasting activity).

Because what one person sees as a warning may not be viewed as a warning by others, several basic questions must be addressed while an EWS is being developed or when its effectiveness is being evaluated. Do people agree on what is meant by "early"? What constitutes a "warning", exactly? Who is to be warned? Does everyone (e.g. the government, the media, the public, the military) need to be warned at the same time, in the same way and by the same warning system?

Other important questions arise. How is a warning affected by the reliability and credibility of the sources of information on which it is based? Are there levels or degrees of warning? For instance, is the warning to be like a "go-no go" or "on-off" warning switch (e.g. high alert or no alert), or are there to be levels of warning like traffic lights: green, yellow and red? Or should the levels of warning be graded like those used for weather phenomena such as tornadoes and hurricanes (e.g. a watch, a warning and an alert)?

There are EWSs for just about every conceivable process or hazard of concern to individuals, societies, corporations and governments. They exist on the global scale (e.g. monitoring global warming and stratospheric ozone depletion), at the national level (e.g. regional drought, desertification, famine and large-scale flooding) and at the local level (e.g. infectious disease outbreaks, nutritional changes and flash floods). They are also constantly being created for newly identified threats (e.g. terrorism, West Nile virus, SARS), being revised for changes in existing threats (e.g. food insecurity, invasive species) and being critiqued for shortcomings (just about every system).

For the most part, EWSs are under constant scrutiny, with each half-generation (every 10 years or so) trying to develop the perfect system. Indeed, numerous examples could be cited of truly successful systems that have been credited with the saving of lives and livelihoods and the protection of property. The task, however, is daunting. People and societies have always been in conflict with a varying or changing climate, and existing warning systems are incessantly being challenged by nature in general and by variations in the climate system specifically. Making the task of early warning even more difficult is the fact that societies, let alone ambient environmental conditions, are themselves constantly changing. What this means, of course, is that both the quick-onset event and the slow-onset ("creeping") process to be warned about are always embedded in a context of other compounding events and processes, and this synergy frequently leads to what are called "complex humanitarian crises". Complexity, however, should not be used as an excuse for inaction.

Some tools of the early warning trade

A range of tools exist to help provide warnings about events that could threaten the stability of a given society, whether that society is technologically complex or not. For example, meteorological and hydrological services have used advanced technologies in EWSs for decades to warn about all kinds of extreme hydro-meteorological events. In addition, woven into the traditions of indigenous cultures around the world are warning systems for similar events that, though not as technologically advanced, have proven effective for generations in warning local populations about impending threats. Regardless of technology, these warnings – as well as those for a wide range of other events – utilize appropriate tools to mitigate and adapt to threat impacts, improving the likelihood that hazards will be overcome. Four such tools are indigenous knowledge, Geographic Information Systems (GIS), remote sensing and forecast warning terminology.

Indigenous knowledge

Gregory Pierce

Indigenous knowledge comprises the manners and customs that are cultivated intergenerationally by a group of people who have an intricate awareness of their own local environment. Once widely denigrated by Western researchers and colonial agents as the primitive practices of primitive cultures, the importance of indigenous knowledge has been increasingly recognized by researchers over the last decades, especially in terms of its benefits for disaster risk reduction and early warning. Its value specifically derives from its ongoing development as a shared resource that is passed informally from generation to generation, enabling adaptation *within* a community to variations in local environmental conditions; rarely overtly emphasized, indigenous knowledge embodies the traditions that are woven into the very life of a culture evolving through time, the cultural "memory" that ensures its survival from one generation to the next (International Strategy for Disaster Reduction, 2008).

Interestingly, because of its cumulative nature, indigenous knowledge evolved for local survival often also acts as an adaptive mechanism that mitigates the development of what could otherwise prove a creeping threat. In the Turpan Depression of Xinjiang Province in the arid west of China, for example, the *karez* irrigation systems have delivered life for over 2,000 years in one of the most hostile climates on earth. Using mainly hand-tools and an accumulated understanding of the porous rock and slopes of the landscape, traditional groups in this harsh environment maintain thousands of kilometres of underground channels into which snowmelt from the Tianshen Mountains flows, sheltered from the threat of evaporation in up to 50°C desert heat. Continual refinement of indigenous methodologies for constructing and maintaining the systems – development of regularly spaced vertical wells for ventilation, orientation and repair access to the channels, for instance – ensures the provision of stable, gravity-fed outflow and high water quality that has enabled the people in this otherwise barren region to flourish agriculturally for millennia.

Significantly, practices and policies that have over decades and centuries been effective in one community not only encourage participation by members of that community as empowered stakeholders in disaster reduction efforts, but have also been successfully transferred to other communities facing similar threats. Furthermore, development projects planned by outside agencies consistently prove more effective when traditional methods are consulted as valuable complements to project design and implementation. Ironically, the informal, once-derided means by which indigenous knowledge is passed on are today considered by more formally trained, technologically inclined disaster planners as a valuable paradigm for disaster education programmes around the world (ibid.).

Indigenous knowledge is now recognized as an essential tool in developing mitigation and adaptation strategies for the threats of the twenty-first century. In fact, governments, the United Nations and NGOs around the world are working cooperatively with indigenous groups to understand and incorporate many aspects of local knowledge into the development of wider-ranging disaster mitigation and early warning systems. Where practical, these agencies are also providing assistance to indigenous groups by informing them about the future threats they will face, such as climate change, and by encouraging them to integrate appropriate technologies – not to replace but to reinforce traditional methods and customs – to mitigate the impacts of such new threats for which they could not have a cultural "memory".

Geographic Information Systems (GIS)

Jennifer Boehnert

Geographic Information Systems (GIS) is a technology designed to collect, store, analyse and display data regarding real-world objects (e.g. rivers, cities, etc.) and events (e.g. hurricanes, droughts, etc.). The "G" in GIS is the unique aspect of this technology; it represents the association of geographic data that can be located on the earth's surface (spatial data) with data that contain descriptive information about these real-world objects and events (attribute data). Through visualization, a comprehensive understanding of these events, along with event-specific impacts, threats and outcomes, can be achieved. The ability to overlay disparate datasets with different themes is another important aspect of GIS technology (fig. 1.1). For example, GIS can be used to ask questions such as which areas are below sea level, have a population density per square kilometre greater than 900 and are susceptible to frequent flooding. These types of questions – and their answers – can greatly improve our understanding of real-world events and their effects. But although computers can help to visualize and analyse the data, by themselves they cannot solve complex problems; human intervention and interpretation are needed. The

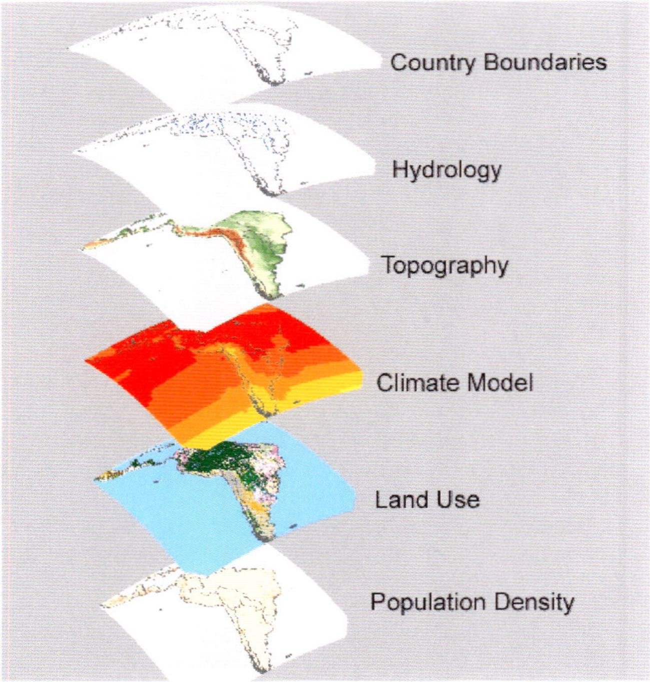

Figure 1.1 A GIS overlay combining several different physical and political features
Source: J. Boehnert.

human component in GIS, along with data analysis and visualization, makes GIS an important tool for decision-making and monitoring natural hazards.

GIS can be used to make a baseline assessment of the vulnerability of a population or evaluate the risk to a geographic region resulting from a particular hazard. Vulnerability mapping can be used to facilitate the development of an emergency plan and better identify preparedness and response actions in case of a hazard. For example, rainfall and wind thresholds can be monitored using GIS technology. During the monitoring phase, GIS can be used to acquire, fuse and analyse a large amount of real-time weather data quickly. Once data are mapped in a GIS, they can be provided as critical guidance to decision-makers, which can aid the mitigation, planning, response and recovery processes. Furthermore, if predetermined thresholds of risk levels are approached, the GIS system can be used to alert emergency managers and first responders to the threat and quickly identify vulnerable populations.

GIS is already being employed in several EWSs around the world. A good example is the UN Food and Agriculture Organization's Global Information and Early Warning System (FAO/GIEWS). GIEWS has developed a suite of software tools called the GIEWS Workstation, a one-stop, web-based system providing maps, tables, charts and documents on food-security-related issues. GIS is a central part of GIEWS's web workstation, providing access to information based on spatial locations as well as spatially related information such as real-time satellite images and base data. GIEWS Workstation facilitates access to food-security-related information for monitoring food supply and demand in virtually all countries in the world.

Another illustration of the use of GIS is the Mapping Malaria Risk in Africa (MARA/ARMA) initiative. This is a collaborative project that collects data from Africa-based regional centres to map the risk of malaria on the continent. GIS is used extensively in this project to collect and store data as well as to generate maps to illustrate areas at risk. Again, this critical information is disseminated to decision-makers in an effort to ensure an effective response to potential outbreaks. It is quickly becoming the pre-eminent data resource for malaria risk mapping in Africa (MARA, 2004).

With enhanced interplay between data formats and improved access to data through the Internet, GIS can become an integral part of just about any early warning system.

Remote sensing
Stefanie Herrmann

Numerous satellites (e.g. NOAA AVHRR, Terra MODIS, Meteosat, GOES) with sensors operating in different portions of the electromagnetic spectrum are detecting energy reflected or emitted globally from the earth's surface and atmosphere day and night. Acquiring such information about the earth's surface from space by making use of the interactions between electromagnetic energy, the atmosphere and the earth's surface is called remote sensing, and it has been an important source of early warning information for several decades. It also constitutes a major input into GIS.

Remote sensing data acquired at high temporal frequency have a great potential for real-time monitoring of cloud movements, sea and land surface temperatures and vegetation conditions, producing data with multiple applications in early warning. With these data, for example, the formation and paths of tropical storms, winter storms and sandstorms can be monitored, and specific warnings about location, time and intensity of these high-impact weather events can be issued hours ahead of actual impact. Early warning of slow-onset phenomena, such as droughts and floods, also benefits from remote sensing. For instance, the long-term monitoring of sea surface temperatures, which in turn affect large-scale atmospheric dynamics and precipitation as well as vegetation conditions, is used to assess drought and flood hazards. Remotely sensed vegetation monitoring coupled with weather predictions form the basis for fire hazard warnings. Furthermore, early warning systems for climate-linked disease outbreaks, such as cholera and malaria, use remotely sensed data to infer the presence of disease-causing agents by monitoring water temperatures and nutrient concentrations, which have been found to correlate with disease outbreaks. Physical and biological changes in oceans are also monitored from space.

The vantage point of space, particularly when combined with the results of field-based vulnerability assessments in a GIS, has proven remote sensing to be a valuable tool for the early warning and disaster management communities, and is increasingly being applied by governments and non-governmental organizations to decrease the impacts of climate-, water- and weather-related hazards.

Forecast warning terminology

All meteorological and hydrological services, as well as many climate-related organizations, have their own specific terminology for warning their constituents (e.g. governments, corporations, private citizens) about climate-, water- and weather-related hazards. As an example, the US National Weather Service (NWS) has developed terminology for weather- and climate-related warnings.

- *Outlook:* The potential for a hazard exists, though the exact timing and severity are uncertain.

- *Watch:* Conditions are favourable for the occurrence (development or movement) of a hazard. The public should stay alert.

- *Warning:* A hazardous event is occurring or is imminent. The public should take immediate protective action.

- *Advisory:* An event that is occurring or imminent is less severe than one that merits a warning. It may cause inconvenience, but it is not expected to be life- or property-threatening if normal precautions are taken.

- *Alert:* The highest level in the warning system, issued when a hazard is sure to occur.

- *Statement:* Detailed follow-up information to warnings, advisories, watches and outlooks is provided.

Readers must keep in mind that for the most part early warning systems differ from country to country. Some governments consider production and issuance of a warning as a stand-alone activity, leaving responsibility for interpretation and response to other agencies. Others look at EWSs in a more holistic way, believing they encompass production and issuance of warnings plus responses to impacts.

Early warning terminology can also vary from country to country, and from one type of hazard to another even within the same country. The entries in this book are drawn from various countries; individual readers can compare these examples to their own society's threat terminologies.

EWS quick facts

- EWSs are politically sensitive; they are not politically neutral.
- One EWS does not fit the needs of all users; one EWS does not fit all hazards.
- People debate the meaning of each word in EWS (early – warning – system).
- Early warning is a high-visibility, risky and thankless job.
- "Success has many fathers. Failure is an orphan."
- People tend to recall failed warnings more often than successful ones.

- EWSs are integral to sustainable development.
- People have different views about the structure and function of an EWS.
  - EWS as a technical unit issuing only quantitatively based warnings.
  - EWS as a holistic method including warning, communication and response.
- Not all governments or EWSs want transparency in early warning processes.
- No agreement exists on how best to express “likelihood” or “probability”.
- There are unknowable and knowable surprises.
- There should be degrees of warnings:
  - Outlook
  - Watch (earliest warning)
  - Warning (earlier warning)
  - Alert (actual warning)
  - Hotspot (target of threat identified)
  - Flashpoint (threat is imminent; little lead-time exists).

Threats: The perils that compel early warning

Every society in the world has its own unique set of threats about which to be concerned and for which to prepare. Some of these threats are natural, while others are directly or indirectly human made. Some can be well prepared for, while others cannot be. Regardless, threats of one kind or another will always exist. While we cannot necessarily eradicate all of them, we can become better aware of their likelihood and potential impacts. For most of these threats, both perceived and real, an early warning apparatus, however formal or informal, has been established. Although some of these warnings may not be labelled as “warnings”, they are, in essence, warnings based on forecasts, projections, scenarios and trends as a result of tracking a selected, explicitly identified set of indicators.

In the end, every society, whether rich or poor, industrial or agrarian, democratic or totalitarian, capitalist or socialist, must determine which threats are credible and based on reliable indicators (qualitative as well as quantitative) and which ones are based on less reliable factors, such as unsubstantiated rumours. The difficulty is providing a reliable warning of a potential threat with enough lead-time for recipients of the warning to take appropriate evasive action. In many respects, consequently, the problem with early warning boils down to the common difficulty of perception versus reality.

The perception of the role of EWSs

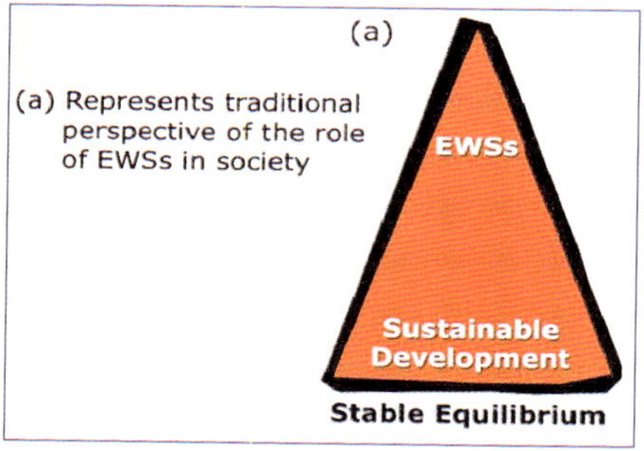

Figure 1.2 Traditional perspective of the role of EWSs in society

Figure 1.2 is a graphic representation of how governments tend to look at EWSs of all kinds, including those for climate, water and weather. The pyramid suggests that governments view societies as stable entities resting on firm foundations.

At its apex is a proverbial "searchlight" that clearly illuminates the perimeter around the base of the pyramid. When a troubling situation enters the illuminated area near the base, a government (in theory at least) initiates a response mechanism to resolve the problem and bring life back to normal. This perception of the role of EWSs in society is, however, readily challenged.

Recent natural and human-made events do not support this pyramid view of societal stability and the role of EWSs. In the United States, for example, the 9/11 terrorist attacks on the Twin Towers and the Pentagon (2001) and the impacts of Hurricane Katrina (August 2005) both highlight how the stability of a society can be shaken by poor implementation of or ineffective early warning systems. Similar examples of destabilizing events occurring because of poor foresight about the importance of effective EWSs can be found in many countries.

Clearly, societies need to rethink how they view, maintain and use their early warning systems if they want to ensure that these mechanisms are effective when needed.

The reality of the role of EWSs

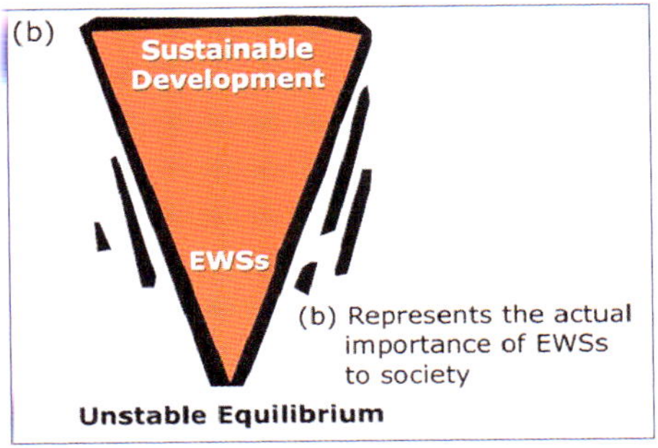

Figure 1.3 Actual importance of EWSs to society

In fact, the reality of EWSs in society can be portrayed graphically as a pyramid resting on its apex in a position of unstable equilibrium, as in figure 1.3. Societies and the governments that lead them are actually dependent on a wide array of early warning systems focused not only on terrorism but also on economics, politics, natural resource use, environmental changes and natural hazards. The better and more effective these EWSs, the flatter the apex becomes upon which the pyramid rests. While societies and governments can never be completely stable, improved warning systems go a long way to increasing social stability.

The situation unfailingly returns to the problem of sustainability. Immediately after a crisis, an EWS receives a lot of attention. As time passes in the absence of similar crises, however, interest in and funding for the EWS sharply decline. But EWSs are much more important than governments often realize; the reality, as illustrated in figure 1.3, is that social, political and economic stability depend on them. Without effective EWSs, a society is highly vulnerable to natural and human-induced threats. At a general level, all societies are aware of this. In practice, however, intervening factors detract from the efficiency of EWSs, leaving societies increasingly vulnerable to the adverse impacts of numerous kinds of hazards. Ultimately, one of the central goals of a government should be to flatten the apex of the inverted pyramid that is its society by improving moral and financial support for its early warning systems in order to build more stability into its social, political and economic systems, which always exist in tenuous equilibrium.

Surprises

Unstable equilibrium obviously does not bode well for a goal of stability. The problem, of course, arises from the element of surprise – the fact that not every possible circumstance can be adequately planned for. Indeed, central to the very definition of surprise is the notion of the unexpected; yet many people use the word "surprise" independently of this notion. For example, at some time or other everyone has uttered such expressions as "I was semi-surprised", "almost surprised", "hardly surprised", "a little surprised", "somewhat surprised", "sort of surprised" and so forth.

The truth is that there are "knowable surprises", even though this expression seems contradictory. Of course, if you consider only denotations, you can in no way know that a surprise is coming. But there are surprises that resonate beyond the dictionary's definition and are, essentially, knowable, as in the example of hurricanes: although the timing, magnitude and exact landfall of a specific event may be unknown, the fact that certain areas are prone to hurricanes is known. Despite the obstacles presented by these unknowns, that the knowable component exists should facilitate the development and implementation of mitigation strategies to counter the threat. Indeed, these knowable components of surprises are why governments and donors must take early warning activities much more seriously than they have in the past.

It is not enough, however, for a society to have an accurate forecast of an impending hazard: a good forecast system is only one part of a broader early warning system in which effective communication of the warning and response mechanisms are integral parts. Having a poorly supported early warning system is also not good enough. If it has only low levels of moral and/or financial support, a system could become ineffective in delivering appropriate warnings to those in need because it could fail to provide them with enough time to respond proactively to those warnings.

Effective early warning systems, on the other hand, can empower governments in ways that help them in the face of climate-, water- and weather-related hazards to protect their citizens as well as their political stability. Several examples exist where poor responses to forecasts and impacts of climate-, water- and weather-related hazards have contributed to the downfall of governments – as was the case during the 1968–1973 drought in the West African Sahel when three governments were overthrown by drought-related military coups. In 1974 the Ethio-

Figure 1.4 Cover of EEA publication, *Late Lessons from Early Warnings: The Precautionary Principle 1896–2000*

pian government was also overthrown, with drought again serving as the justification by the military junta.

Those involved in environmental monitoring activities have at one time or another been faced with attempts to reduce their funding, especially if the hazard they are monitoring has not manifested itself for some time. To minimize unexpected surprises, however, this "benign neglect" of EWSs by governments must be made explicit and avoided.

The future has been arriving earlier than predicted; take global warming, for example

With each passing decade since the 1970s, it seems that more and more of the signs scientists have warned us about related to the gradual warming of the earth's atmosphere have been observed. Many of these signs are emerging far more quickly than originally predicted, yielding an increasingly palpable sense of urgency for response planning. For example, sea level continues to rise; 96 per cent of the world's glaciers are receding; warm climate ecosystems are moving upslope to higher altitudes and into previously cooler climates and latitudes; and exotic species and disease vectors are appearing in new locations poleward, adjusting to warmer winters and hotter summers. Droughts seem to be recurring with greater frequency and intensity in some drought-prone locations, and floods are doing the same in flood-prone locations. Additionally, Arctic sea ice has been disappearing at an increased rate, drawing nearer and nearer every year to a dangerous tipping point beyond which an ice-free Arctic will no longer be avoidable.

Many of these changes are taking place earlier than expected, at rates faster than expected and in places where they were often unexpected. Furthermore, we are starting to see stronger storms, some of which are now being labelled "superstorms". In fact, we are witnessing the appearance of "seasons of superstorms".

While many of the computer-model-based climate change scenarios yield the foreseeable consequences of global warming out to the year 2050, 2070 or even 2100, we are now already witnessing in different parts of the globe some effects scientists had expected to take place in the distant future. Coral reefs are dying worldwide; permafrost is melting; each new year seems to be ranked in the hottest 10 years on record; tropical storms in the Atlantic, Pacific and Indian Oceans are increasing in frequency and intensity; and so forth. These are changes that had been suggested verbally and in print for several decades, but now they are no longer speculative. Now, they are real.

Compounding the physical and biological changes accompanying global warming (all observers admit that the climate has warmed by about 0.7°C since the early 1900s) are demographic changes, such as in population growth and migration, land transformation and land-use pattern shifts, increasing exploitation of a wide range of natural resources and intensifying water and food shortages. In addition there has been a growing trend worldwide of population movement towards the coastal areas, which are going to be increasingly at risk from tropical storms, storm surges and sea-level rise. More than ever, timely warnings for the entire range of potential threats – old, new and intensified – are essential in response to the changing dynamics of the environment.

Strengths, Weaknesses, Opportunities and Constraints assessment

The UN Development Programme (UNDP) uses an assessment technique called SWOC (Strengths, Weaknesses, Opportunities and Constraints) to evaluate various programmes and activities. The following paragraphs apply SWOC to early warning systems in general, especially emphasizing early warnings related to atmospheric processes such as droughts, floods, frosts, fires, heat and infectious disease outbreaks. Important to keep in mind is that each weakness or constraint can also be viewed both as an opportunity for future improvement and as an evolving challenge to society.

Strengths

People want to know what the future has in store for themselves, their families and their societies. The same is true for every government. A glimpse into the future, therefore, no matter how reliable (or unreliable), is always viewed as having some value to someone, so the strength of an early warning about any threat that might impinge on routines or livelihoods or the workings of government is seen as having value. Value, however, is a loaded term, because what one person values another may not; what one government values, another may not.

Early warning implies that there will be "ample time" to react to the information conveyed by the warning. Ample time refers to lead-time and is a relative term, dependent on the hazard or concern. An early warning several months in advance of the development of an El Niño episode in the tropical Pacific can provide many countries with several months of usable lead-time. A tornado warning, on the other hand, may be issued only a matter of a few minutes or hours ahead of a tornado's impact. Both, however, can provide enough lead-time to take evasive, protective action.

Nevertheless, there are time constraints on warnings. To hear about a flash flood as the water is rushing at your doorstep does not provide ample time to act – the warning is too late to be usable. Also not very useful is an early warning that is given too far in advance of the impact of what is being warned about; for example, if you are living in a region you know is prone to flash flooding, can that knowledge be considered an early warning? In both of these flood scenarios, the information provided might be of some use, but does not, by itself, maximize the potential benefits for those being warned. The former example may allow you to save some possessions and even your life; the latter, on the other hand would only enable you to protect yourself against a low-probability, worst-case, flash-flood scenario.

The bottom line about the value (strength) of early warning systems is that governments, the media and the general public, as well as those most at risk to climate-related or other hazards of concern, can be warned to take appropriate action. They must only be receptive to being warned.

Weaknesses

In theory and on paper, early warning systems function quite efficiently. In reality, however, they seldom work smoothly. Despite the best efforts of individuals, agencies and governments, neither nature nor the climate follows a set pattern or a government's best hopes for their good behaviour. In planning for hazards, therefore, as American ecologist Barry Commoner once said of the natural environment, "You can't change just one thing." This is especially true as far as human interactions with climate systems are concerned, where demographic changes (population shifts as well as increasing affluence or increasing poverty) over time alter the impacts on societies and ecosystems of constantly varying and changing atmospheric behaviour.

Thus other factors beyond the issuance of a warning itself – such as the ability of those at risk or agencies responsible for them to receive and understand the warning and the phenomenon being warned about, political access of at-risk populations, vulnerability of communications and transportation infrastructures to hazard impacts, or even the desire or capacity of a government to take appropriate action – influence the actual, as opposed to the hypothetical, value of an EWS.

Needless to say, there are weaknesses inherent in early warning systems, too. For example, usually no single indicator can provide the optimal lead-time for a warning. Different groups with different perspectives and interests are likely to have chosen their own sets of indicators for local hazards that they have had to face in their lifetimes. How then are reviews and assessments of all possible indicators to be taken from all the various sources in order to create a reliable, credible and timely warning of an impending threat or hazard?

Despite such questions, early warning systems do require set structures as well as explicit functions (or purposes). In reality, they are often bureaucratic units embedded within larger bureaucratic units which are themselves embedded within even larger bureaucratic units. Considering this reality, EWSs require flexibility because reporting procedures across these units can be designed either to accelerate or to inhibit the flow of warning information. Unfortunately, bureaucracies are often run by people who cannot help but have vested interests in their organizational units, if not their work; these people collect information selectively, focusing on their unit's specific area of jurisdiction or the realm of their own experience or expertise. This is normal human behaviour. As a consequence, however, bureaucratic jurisdictions often become obstacles to the quick issuance of (or the response to) an early warning.

Individual perceptions about the strength or weakness of certain indicators in terms of predictive power can also hinder a quick response to an early warning. For example, in the late 1980s in Ethiopia, even though the country's National Meteorological Agency (NMA) had issued warnings of the likelihood of meteorological drought, the Minister of Agriculture, noting that *his ministry's* forecasters had not come up with the same drought scenario, challenged those forecasts instead of treating them as early warnings. Because chronic hunger had already become a fact of life and food assistance had to be requested and brought into the country, who was correct is difficult to determine. What is important is that the meteorological service did provide the government with a "heads up" on a drought, a food crisis and the potential for famine, and, despite the agriculture minister's challenge, the government did eventually have to seek humanitarian food aid.

Many affected governments and donor agencies want to be the first to receive a warning, often in private, either to avoid causing panic among the public or for other socio-economic or political reasons. They finance the development and maintenance of early warning activities because they do not want to be surprised by the sudden emergence of a threat about which they should and could have been forewarned.

In the end, information is power, and information received by one special-interest group, a ministry, for example, ahead of others can be politically very rewarding. This truism, however, suggests another weakness in early warning systems, because communities or groups at risk are ultimately the ones that should be warned as soon as possible to provide them with as much lead-time as necessary to mitigate the effects of a hazard; however, at-risk people often do not have the resources needed to cope with a forecasted hazard. Consequently, governments, which possess more resources to respond to foreseeable hazards than individuals or groups, often need to be informed first.

Opportunities

Opportunities for developing new or improving existing EWSs are, in fact, limitless. Indeed, many proverbs exist in societies worldwide about being precautionary, so there is little need to sell a precautionary level of awareness about EWSs to governments; however, problems often emerge when finances are needed, especially for maintaining an environmental monitoring capability. This reality becomes increasingly apparent as EWSs are considered for phenomena that are infrequent or rare.

New technologies for and new approaches to early warning have provided new opportunities for reducing the risk of climate-, water- and weather-related problems. Such opportunities are provided by improved scientific understanding of the hazards of concern; by new technologies for detection purposes; by better identification and targeting of at-risk populations; by improved understanding of second-order, hazard-related impacts; and by a seamless transfer of information to those who are in positions to respond to warnings.

Communication advancements related to the issuance of warnings and the targeting of those warnings to relevant agencies and at-risk populations have also greatly improved. The Internet, satellite radios and mobile phones now serve as vehicles for warning dissemination. Satellites, for example, provide images with improved resolution, and Google Earth on the Internet provides images of impacted regions and localities in real time, enabling those with the requisite electronic devices (e.g. radio, TV, computer, PDA, wireless phone) to monitor the progression of a hazard.

Lastly, the globalization of communication and education has worked wonders by bringing stories of both successes and failures to the attention of those facing similar risks and threats, whether they happen to be in urban centres or remote areas, in other parts of the globe.

Constraints

Not everyone in a known at-risk location has equal access to early warnings, regardless of the means used by those responsible for early warning systems to broadcast them.

For example, even though connectivity to EWS output is crucial to reducing impacts, at-risk populations are not always connected to various types of electronic media. Because of this reality, other means of delivery need to be made routinely available, such as the use of satellite radio (e.g. RANET), mobile phones and public sirens. Language differences are yet another potential constraint. There is a need to make sure that warnings are provided in appropriate languages – foreign, non-scientific and even non-verbal. No single communications medium, however, will ever reach an entire at-risk population.

In the absence of a hazard over an extended period of time (which varies by threat, country or group), a tendency exists to downplay the continued need to finance the formal structures of the early warning systems associated with that hazard. The costs are often viewed as not worthwhile, given the relative infrequency of the hazard. What comes to mind is the TV commercial about an appliance repair man who sits around day after day, bored as could be, waiting for a call to repair an appliance that has been manufactured so well that it never needs to be repaired. Eventually, however, he will receive the call; similarly, an EWS will eventually prove invaluable when a threat inevitably does arise.

Another problem occurs when a warning has been issued but the hazard fails to occur, or, on the other hand, when hazards prove challenging to predict and warn about because they are either quick onset or of varying intensities. As a result, belief in the effectiveness of early warning systems is likely to decrease. Such a reaction is, however, inappropriate. Any forecast of a potential hazard has a possibility of being incorrect (the hazard could weaken or change location, or its impacts might not be as severe as expected because appropriate actions were taken to mitigate them, among other possible reasons).

As problematic as is the issuance of infrequent early warnings in proportion to the frequency of a certain hazard's likely occurrence, the problem of issuing warnings too often must also be considered. This scenario can generate the "cry wolf" problem when a real emergency arises and a warning is not taken seriously (fig. 1.5). For example, once when the US Department of Homeland Security in Washington, DC, elevated the terror alert from Yellow to Orange, a local official in the US state of Arizona referred to the warning as "Orange-lite", suggesting that one warning for the whole country was inappropriate as the likelihood of a terrorist attack in his district was considerably lower than it would be for a major urban area in, say, the eastern United States.

Figure 1.5 The dangers of crying "wolf"

Furthermore, warnings, forecasts and projections are constantly being issued about this or that hazard to various parts of society. This profusion of forecasts and warnings, issued by a range of institutions, agencies and groups and often suggesting conflicting levels of threat, can confuse the public and the media and decrease the likelihood of either taking a warning seriously. Such a situation also makes it difficult to know who to rely on for early warning of an impending climate-, water- or weather-related threat. Because preventing these erroneous or misinformed pronouncements about threats is not possible in most cases, however, a consistent approach must be developed to inform the public about the reliability of specific forecasts issued for the same threat. At the same time, as broad an audience as possible must be empowered to make informed decisions when faced with the anticipated consequences of possible threats, even if those threats sometimes fail to materialize.

Accordingly, this book is intended as a first step in making early warnings of potential threats to the environment and to society more credible, reliable, usable and useful to a broad audience. In realizing this intention, the following chapters gather together for the first time essential material from across the sizeable but largely scattered body of academic work that exists on early warning systems, presenting significant concepts from the field in language accessible to readers from all backgrounds, from students to community organizers to activists to government officials. In this way, it is hoped that this book advances the vital idea of "usable science" in the twenty-first century.

2

Climate-, water- and weather-related hazards

In the sections that follow, a selection of climate-, water- and weather-related early warning systems are presented. Each entry introduces the actual levels of warning or the terms that forecasters use to inform governments and the public about a particular hazard's level of threat. Some of these systems use words to identify threat levels, while others use numbers and still others use colours. Included in the sections are photographs or graphics related to hazards for which early warnings are issued. Accompanying pages provide descriptions, written by contributing authors, related to specific hazards.

Brief cases are embedded throughout the text to highlight the need, use, value or failure of various EWSs in real-world situations. These cases are meant to be illustrative, and do not represent the hundreds of early warning systems, both formal and informal, that exist around the globe. The hope is that readers will see just how important early warning systems are and how important it is for governments and organizations to maintain them, even in the absence of a constant threat from hazards that appear infrequently.

Hurricanes

Roger Pielke, Jr.

- *Tropical Storm Warning:* A warning that sustained winds within the range of 34–63 kt (39–73 mph or 63–118 km/hr) associated with a tropical cyclone are expected in a specified coastal area within 24 hours or less.

Heads Up! Early Warning Systems for Climate-, Water- and Weather-Related Hazards, Michael H. Glantz (ed.), United Nations University Press, 2009, ISBN 978-92-808-1169-8

- *Hurricane Local Statement:* A public release prepared by a local National Weather Service office in or near a threatened area giving specific details for that area on (1) weather conditions, (2) evacuation decisions made by local officials and (3) other precautions necessary to protect life and property.

- *Hurricane Watch:* An announcement to specific coastal areas that hurricane conditions are possible within 36 hours.

- *Hurricane Warning:* An alert indicating that sustained winds of 64 kt (74 mph or 119 km/hr) or higher associated with a hurricane are expected in a specific coastal area in 24 hours or less. A Hurricane Warning can remain in effect even though winds may be less than hurricane force when dangerously high water or a combination of dangerously high water and exceptionally high waves are present in the warning area.

(CPC, 2006)

Twice a year, in May and August, the US Climate Prediction Center (CPC) posts the *Hurricane Outlook Discussion*, which analyses climate influences on and expected hurricane activity in the Atlantic and Caribbean Basins from June to November.

The word hurricane derives from the Spanish *huracán*, which originally came from the dialects of peoples indigenous to the Caribbean and Latin America. Historical records of hurricane events, however, begin only after the European conquest of North America. Tropical cyclones today are named once they have reached tropical storm strength, and are considered hurricanes once they exceed a wind speed of 119 km/hr (74 mph). Thus, "a hurricane is a severe tropical storm that forms in the North Atlantic Ocean, the Northeast Pacific Ocean east of the dateline, or the South Pacific Ocean east of 160°E... In other regions of the world, these storms have different names: Typhoon (in the North-

Figure 2.1 Hurricane Dennis (1999) caused severe beach erosion along the Outer Banks of North Carolina, US. The pounding surf and surge undermined and claimed the pictured beach home, while leaving other homes unscathed along Virginia Dare Trail in Kitty Hawk
Source: Dave Gatley/FEMA.

west Pacific west of the dateline), severe tropical cyclone (the Southwest Pacific west of 160°E or the Southeast Indian Ocean east of 90°E), severe cyclonic storm (the North Indian Ocean), tropical cyclone (the Southwest Indian Ocean)" (Pielke and Pielke, 2003).

At the end of the twentieth century, Pielke and Pielke (1997: 182–191) identified several lessons related to hurricanes in the North Atlantic (several of these lessons apply to cyclones and typhoons as well):

- Tropical cyclones are the most costly natural disaster in the United States and worldwide.
- Tropical cyclone forecasts (seasonal, intensity and track) can continue to improve, though societal benefits associated with improved forecasts depend on their effective use.
- Climate varies on all measurable time scales.
- There is considerable reason to prepare better for hurricanes independent of concerns about global warming's impacts on the frequency and intensity of hurricanes.
- Tropical cyclone landfalls highlight existing levels of societal preparedness.
- When a hurricane approaches a particular stretch of coastline, effective decision-making depends on the existence of plans, procedures and prior preparations for the event.
- Better knowledge of hurricanes, by itself, is generally not sufficient for changing the public's behaviour.
- Although the processes of preparedness for hurricanes are generally well understood, how actually to translate that knowledge into lowered vulnerability or increased resilience is still neither well understood nor well implemented.

Figure 2.2 The 1998 hurricane season featured the strongest October hurricane on record, Hurricane Mitch, which claimed more than 11,000 lives in Central America. The last time a single Atlantic hurricane caused so many deaths was in 1780

Source: Satellite image, NOAA.

Figure 2.3 The National Weather Service's Tropical Prediction Center predicted the path of Hurricane Katrina on 26 August 2005
Source: NOAA.

Hurricane Katrina: Good forecast, poor response
Michael Glantz

Hurricane Katrina reached Category 5 status in the middle of the Gulf of Mexico and still retained Category 4 strength by the time it made landfall in the US Gulf of Mexico states of Louisiana, Mississippi and Alabama on 29 August 2005. It was called a massive hurricane, a top-strength storm, an incredibly strong storm and a superstorm, and its intensity and trajectory had been well forecasted for about 60 hours in advance of landfall (fig. 2.3).

Many scenarios over the years had projected the possible negative impacts of a Category 3, 4 or 5 hurricane hitting New Orleans head on, and in fact Katrina did most of its damage after it made landfall, when its strength had dropped to that of a Category 3 storm (winds of 178–209 km/hr or 111–130 mph).

Such a natural hazard was sure to bring about some level of death and destruction; however, the excessive damage from the event was significantly greater than even the experts had predicted. Much of the blame for the extreme mortality (more than 1,800 dead), destruction and misery caused by Katrina rests with the inadequate political response to the adverse impacts of the event. For example, the urban poor were known to live in high-risk locations relative to the pathway of known natural hazards in the region, but evacuation plans apparently failed to utilize these data. Furthermore, the levees that protect the New Orleans area were for at least a decade known to be in need of urgent repair and upgrading. Finally, the federal agencies responsible for responding to such emergencies, most notably the US Federal Emergency Management Agency (FEMA), failed to act in a timely and effective way.

Figure 2.4 Homes flooded by Hurricane Katrina, 31 August 2005
Source: NOAA.

Ironically, as recently as July 2004 a make-believe hurricane, Hurricane Pam, had been devised as a training exercise to engage emergency managers from local to national levels in a scenario where a Category 3 event hits New Orleans. After a week or so of contending with the impacts of hypothetical Pam, a press conference was held to report how successful the agencies had been in their planning for and response to such a scenario. The message to the public was "Don't worry. Your safety is being taken care of." The question, then, is how such disparity could exist between the hypothetical scenario and the actual event?

History shows that tropical storms have barely missed New Orleans on several occasions. An event such as Hurricane Katrina was, therefore, a foreseeable surprise. This foreseeability raises questions about which of the horrendous impacts of this "natural" disaster should truly be blamed on nature and which ones should be blamed on societal – especially political – decision-making. Katrina's destructive power (costing more than US$250 billion) sparked a heated debate about which level of government (local, state or federal) has the responsibility of first responder. A second debate also emerged among scientists about whether Hurricane Katrina, the most costly natural disaster in American history, had resulted from the influence of global warming. Neither of these debates has as yet been resolved.

Significantly, Hurricane Katrina also exposed several disaster-related myths. One myth is about societal preparedness for the impacts of natural hazards. The myth is that affluent, industrialized countries are not as vulnerable as developing countries to variations in climate, water and weather extremes. Another is that the impacts associated with Katrina were the result of a *natural* disaster. Yet another myth perpetuated by political leaders suggests that the poor opted to live in dangerous places (flood-prone areas and exposed landscapes, for instance) around the Gulf Coast. The truth is that they had few options to live elsewhere, given their economic status (Glantz, 2005).

Lastly, a hidden cost associated with Hurricane Katrina remains relatively unaccounted for: the psychological. In the weeks following the event, reporters around the globe wondered how one of the world's most powerful nations seemed for several days to have been incapacitated by and unable to respond effectively to the impacts of a Category 3 hurricane. For their part, Americans were forced to realize that in disaster situations they themselves probably need to be their own first responders.

Cyclone Nargis

Douglas Pattie

Cyclone Nargis ripped through Myanmar's Irrawaddy Delta region as a Category 3 storm on 3 May 2008. Sustaining its strength (Category 2 or above) and moving quickly as it approached land, the cyclone hit the most populous, low-lying area of the delta. Heavy precipitation and a storm surge of over 3 metres, affecting not only the coastal area but up to 40 kilometres inland, washed away hundreds of villages and left over 130,000 dead. Some 2.4 million people were severely affected by the cyclone, and almost 1 million, lacking access to adequate quantities of food and potable water for months following the event, completely lost their livelihoods (Deutsche Presse Agentur, 2009).

The Indian Meteorology Department dispatched an initial advisory about the cyclone to the Myanmar authorities on 26 April 2008, two days before the cyclone actually formed in the Bay of Bengal and nearly a week before it struck land. Updates were provided every three hours, and on 30 April the detailed route, speed and locations of landfall were known.

A press briefing was given to Myanmar's national media on 1 May, and newspaper headlines on 2 May focused on the approaching cyclone. The media broadcast a cyclone alert

Figure 2.5 A farmer and survivor of Cyclone Nargis surveys his flooded farmland, located in the Irrawaddy Delta region along the shores of the Andaman Sea in Myanmar
Source: UN photo by Evan Schneider.

on the afternoon of Friday 2 May when the storm first reached the delta; it swept into Rangoon (population 6 million) early the next day.

As an international commitment through the Economic and Social Commission for Asia and the Pacific (ESCAP) Panel on Tropical Cyclones, advisories are issued four to eight times a day by the WMO Regional Specialized Meteorological Centre (RSMC-New Delhi) to panel member countries when tropical cyclones have formed in the Bay of Bengal or the Arabian Sea. The ESCAP panel countries are Thailand, Myanmar, Bangladesh, India, Pakistan, Sri Lanka, the Maldives and Oman.

Cyclone warnings are issued in two stages. The first-stage warning, known as a *cyclone alert*, is issued 48 hours in advance of the expected development of adverse weather over coastal areas. The second-stage warning, known as a *cyclone warning*, is issued 24 hours in advance of landfall. A pre-cyclone *watch* may be issued prior to the cyclone alert, and a post-landfall *outlook* is issued for areas in the interior that may be affected by the cyclone as it dissipates while moving inland.

Cyclone warnings are disseminated through radio, television, print media, telephones, fax, telex, telegrams and police wireless networks. A specially designed Cyclone Warning Dissemination System, which works via the Indian INSAT satellite, provides area-specific services even when a failure of conventional communication channels occurs. These warnings are issued for the general public, fishermen and farmers; for central and state government officials responsible for disaster mitigation and relief; for industrial users and other establishments located in the coastal areas; and for other stakeholders, including railway, aviation, communications and power authorities. Tragically, in the case of Nargis, warning information was not widely disseminated.

A comparison between Cyclone Nargis and Cyclone Sidr, which struck Bangladesh in 2007, is informative. The much lower death toll in Cyclone Sidr was the result of a far higher level of preparedness. Unlike Myanmar, Bangladesh has a recent history of destructive cyclones, and as a result has developed an effective early warning system and reliable contingency plans for tropical storms, including elevated shelters close to population centres to provide a quick and efficient means of vertical evacuation, the only effective way to escape a storm surge or tsunami.

In the post-Nargis environment, Myanmar's operational meteorological infrastructure needs to be restored and its current meteorological information services enhanced. The urgent need is for a coordinated storm-surge prediction system, better communication and public awareness of warnings, and coordinated disaster preparedness measures. A fundamental technical problem facing Myanmar's Department of Meteorology and Hydrology was the lack of an upper-air observation system and radar network to monitor the progress of the storm. Along with acquisition of such a system, the data-processing and forecasting system including storm-induced flood forecasting must be strengthened, and a back-up electrical power supply is necessary to ensure the availability of minimum operational services under all circumstances. Public weather services should also be enhanced to guarantee better dissemination of weather information and warnings that are clear to all people in affected areas. Furthermore, technical advice for disaster risk reduction, such as risk assessment and risk mapping, will help reduce future vulnerability of people living in the Irrawaddy River Delta.

For the rapidly developing countries that surround the Bay of Bengal, whose peoples and territories have experienced some of the world's most devastating storms, an increasing urgency exists to discover ways to protect themselves from tropical events. After the 2004 Indian Ocean tsunami, for example, a number of initiatives were launched to establish state-of-the-art early warning systems. The lessons learned from the Tsunami Early Warning System implemented by UNESCO's Intergovernmental Oceanographic Commission (IOC) in particular could be useful to Myanmar in its efforts to prevent and mitigate the impacts of future storm surges.

As many Indian Ocean rim countries – especially those that share the resources of the Bay of Bengal – have learned through trial and experience, however, a high-cost system is not a solution by itself because even the most advanced early warning system in the world can only do half the job: alerting governments and other centres of power, such as the military or civil defence, to an impending disaster. The other half requires advanced preparation, which demands public education and political will. The greatest challenge, in fact, is determining how to disseminate warnings effectively in the shortest time possible to large numbers of people who possess considerably diverse communication resources and are spread across vast areas.

People living in coastal communities around the Bay of Bengal have a vital and shared interest, as they are collectively joined by a dynamic and sometimes volatile body of water that is heedless of the intra-border interests of nation-states. Indeed, one lesson of the tragedy of Cyclone Nargis is that sovereignty, isolation and national security are meaningless concepts under impending hazard conditions. Therefore, because the development of an early warning system for storm events is a pure public good undersupplied by the market, the responsibility for establishing such a system falls on the collective shoulders of governments in the region, which must work together to mitigate the damages and protect their communities from future devastation.

Severe winter storms

Stanley Changnon

- *Winter Storm Outlook:* Issued prior to a winter storm watch, an outlook is given when forecasters believe winter storm conditions are possible. It is usually issued three to five days in advance of a winter storm.

- *Winter Storm Watch:* Alerts the public to the possibility of blizzard conditions, heavy snow, heavy freezing rain or heavy sleet. A winter storm watch is usually issued 12 to 48 hours before the beginning of a winter storm.

- *Winter Storm Warning:* Issued when hazardous winter weather in the form of blizzard conditions, heavy snow, heavy freezing rain or heavy sleet is imminent or occurring. A winter storm warning is usually issued 12 to 24 hours before an event is expected to begin.

Severe winter storms occur in the middle and upper latitudes of the globe and are one of the major weather hazards to human life and property and to the environment. They are "high consequence-low probability events": they are infrequent, but, as with major floods and hurricanes, when they occur, major losses follow. In fact, much of the annual loss occurs in one or two exceptional storm events each year. Some winter storms are due to snow, some to ice and some to both, but the worst are typically a mix of heavy snow, high winds, freezing rain and very cold temperatures.

In the United States, the annual number of lives lost in winter storms ranks second only to those lost due to heat waves. Typically, between 120 and 170 persons lose their lives and thousands are injured each year. Property losses due to winter storms are also significant. From 1949 to 2000, for example, losses from snowstorms averaged $280 million a year, with ice storms causing an additional $320 million in losses. Together, their annual average devastation amounts to more than US$0.5 billion. Significantly, losses have been increasing in recent years largely as a result of major damage to large urban areas, which are especially vulnerable to snow and ice storms.

In January 1998 a major ice storm struck the northeastern United States and eastern Canada, causing $6.1 billion in property damages. The prolonged freezing rain associ-

Figure 2.6 A severe winter storm coated most of Nebraska, US, with a thick layer of ice and snow
Source: Nebraska Emergency Management Agency.

ated with the storm ravaged both populated and rural areas, producing the largest insured loss in Canada's history (C$1.44 billion) with 696,590 claims generated. Human suffering in Canada was also extensive, with 28 dead, countless numbers injured and severe and prolonged power outages that lasted for up to 31 days in the bitter cold of mid-winter, affecting 1.6 million utility customers and 4.7 million people. In fact, heavy ice accumulations damaged or destroyed 130 20–50 ton transmission towers and 30,000 distribution poles, wreaking havoc on the entire Canadian electrical transmission system (fig 2.7). In the end, the storm is estimated to have cost Canadians C$3,000 per household.

The agricultural sector and dairy producers suffered further losses that exceeded C$32 million. Agricultural damages included the loss of 700,000 maple trees and damage to 60 per cent of the region's apple trees. Costs for clean-up and replanting were C$76 million, with damage to forests in the Mount Royal area of Quebec alone estimated at C$18 million. Canada's total storm losses were C$5.1 billion.

In the United States, agricultural, forest and recreational areas in many rural locations in northern Maine, New Hampshire, Vermont and New York also sustained severe losses. The storm claimed 17 lives, led to 500 injuries and resulted in power outages of up to 23 days for 546,000 electric utility company customers, affecting a total of 1.5 million people. Some areas of Vermont faced outages that lasted up to 66 days. Notably, the destruction covered a large area, with more than 400 municipalities in Maine, 80 communities in New York and 95 towns in New Hampshire experiencing damages and power outages. In the end, losses insured in the private sector produced 139,650 claims and US$202 million in payments.

The storm blanketed northern New England and New York state with up to 2.5 cm of ice, destroying or badly damaging 17 million of the region's 40 million acres (18 Mha) of forest: 44 per cent of Vermont's public trees were damaged, and the US Forest Service

Figure 2.7 Power and telephone lines sag after a heavy ice storm. Besides disrupting transportation, heavy ice and snow often damage public services
Source: NOAA.

estimated total natural resource losses at over $1 billion. Total losses in the United States from the January 1998 storm were estimated at over US$1.5 billion.

More frequent than ice storms, snowstorms also pose a major weather threat to North America, damaging structures, halting transportation and endangering lives. Long a significant problem for societies, with increased development and greater reliance on modern amenities as well as the not yet fully understood influences of climate change, the frequency of and costs associated with these storms seem to be rising. Indeed, numerous multi-billion-dollar storms have occurred since 1990, including the March 1993 superstorm that caused $6 billion in losses and 285 deaths and the blizzard of 1996 that caused losses of $0.5 billion and claimed 60 lives. In the twenty-first century, snowstorm damages from the east coast storm of March 2001 exceeded US$4 billion. The impacts of three major snow events punctuated the winter of 2002–2003.

Significantly, mounting losses from severe winter weather events since the 1990s have greatly increased interest in the insurance industry for long-term data to assess fluctuations in risk. In fact, important questions that seriously affect the weather insurance industry about both the short- and the long-term impacts of global climate change have been raised in recent years.

Research indicates that the amount of loss from winter storms is somewhat proportional to a society's level of preparation. Local and regional governments that wisely prepare, for example, can reduce vehicular losses with an adequate response system to clear roads. Similarly, individuals can prepare by stocking batteries, flashlights and portable radios so power outages do not cause major problems in their homes, and companies dependent on deliveries of supplies can prepare by increasing inventories in advance of severe weather.

Furthermore, assessments of recent major winter storms in North America have identified three key lessons for future preparations.

- *Society and its institutions are ill prepared for major winter storms.* The sizeable and various damages and the confusion surrounding public and governmental responses collectively reveal that few people know how to respond effectively to the problems created by major winter storms.

- *Changes in society have created new vulnerabilities to winter storms.* Storms over the past 15 years have highlighted the growing vulnerabilities of modern society to major storms. As noted above, these storms have shown how lifestyles and business operations can be seriously disrupted for prolonged periods because of their heavy reliance on electrical power. Even though modern energy resources, water systems, communications networks and transportation facilities tend to be remarkably reliable, they continue to be quite vulnerable to such storms and often in ways unforeseen. The 1998 Canadian ice storm, for example, revealed a need for emergency water systems for drinking water, temporary sewers and hydrants. Furthermore, alternative means of communicating with the public must be planned for because of the ongoing vulnerabilities of modern communications to these storm events.

- *Crucial government and business programmes are missing.* Recent storms have revealed two general actions public and private emergency planners and responders should pursue to enhance their response systems: the need for research on steps to prevent or mitigate risks and the need to develop programmes to create ongoing awareness and better preparedness. Actively working to incorporate these actions into their systems would serve four objectives. They would allow for immediate implementation of specially devised mitigation plans; publicize the roles and responsibilities of government agencies and other stakeholders; enable stakeholders and the public to develop self-protection knowledge; and use print and electronic media to publish and distribute critical disaster-related information. Importantly, these actions would have to be reviewed on a continuing basis to ensure that established plans remain viable in response to societal and climatic shifts.

Superstorm 1993: North America and the Caribbean

American TV's *The Weather Channel* coined the name Superstorm '93 to describe the massive mid-March winter storm system that at one point encompassed 26 of the 48 contiguous United States, part of northern Mexico, Cuba and northeastern Canada. It has also been called "the perfect storm", as three separate systems merged into one massive system along the American coast of the Gulf of Mexico (fig. 2.8). In the US state of Florida it is called the "No-name" storm because inhabitants there are accustomed to named hurricanes that form within a defined hurricane season and make landfall from either the Atlantic or the Gulf of Mexico. This particular storm, on the other hand, developed over the southeastern half of North America. Cuban meteorologists, who successfully forecasted the storm's development five days in advance, recognize this storm as an extraordinary, out-of-season, surprising, extra-tropical event. Other related forecasts were not as good, though they improved as the storm system moved northward along the eastern coast of the United States. Meteorological indicators from the event objectively showed that it was the most severe winter storm witnessed in the United States since the 1950s, causing tremendous impacts on life and property and closing airports throughout the eastern part of the United States for days.

Since Superstorm '93 there has been an increase in the use of the term "super" to describe meteorological events. For example, in 1999 SuperCyclone Orissa struck the Indian state of the same name, which is on the Bay of Bengal and at risk to most cyclones that advance towards Bangladesh. The storm was responsible for more than 20,000 deaths and widespread destruction there. Four years later, South Korea was hit by SuperTyphoon Maemi (3–13 September 2003), which had an enormous negative impact on South Korea's major port city of Busan. Numerous cranes used to load and unload cargo ships were toppled, slowing ship traffic significantly for weeks while new cranes were constructed.

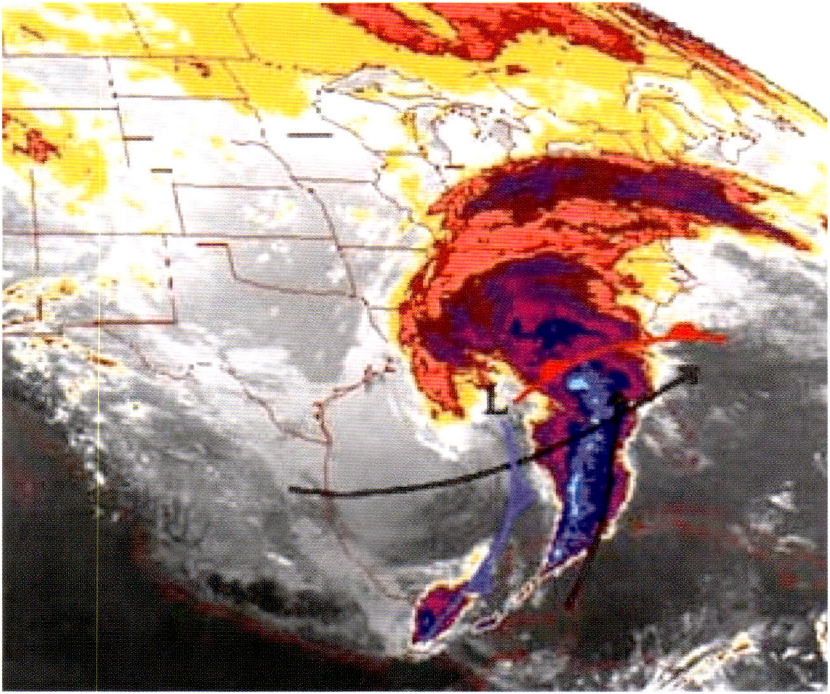

Figure 2.8 The extra-tropical cyclone called Superstorm '93, 13 March 1993, stretched from Cuba to Canada
Source: NOAA National Climatic Data Center.

Are "seasons of superstorms" a possibility?

The recent intensity of tropical storm seasons has raised the spectre of "seasons of superstorms". The 2004 tropical storm season in the Atlantic, for example, saw four hurricanes make landfall in the US state of Florida. The previous record for a season had been three. In the Pacific, 10 typhoons made landfall in Japan that summer (the previous record had been eight). The 2005 tropical storm season was record-setting as well: for the first time there were 28 named storms in the Atlantic, forcing hurricane forecasters to use Greek letters as names for storms.

Such activity leads to worrying questions. Are we witnessing a shift to seasons of superstorms that could accompany global warming? Is this a new possibility for early warning system managers to worry about? Do societies along the foreseeable paths of tropical storm systems – cyclones, hurricanes and typhoons – need to consider developing early warning and response strategies not for a single, one-time super-event in a season, but for several "blockbuster" tropical storm systems each season?

Something new for early warning system managers to worry about... Seasons of superstorms

The 2005 northern hemisphere summertime tropical storm season was:

- the first with 28 named tropical storms
- the first with 15 hurricanes
- the first with four hurricanes classified as Category 5
- the first with four major hurricanes hitting the United States.

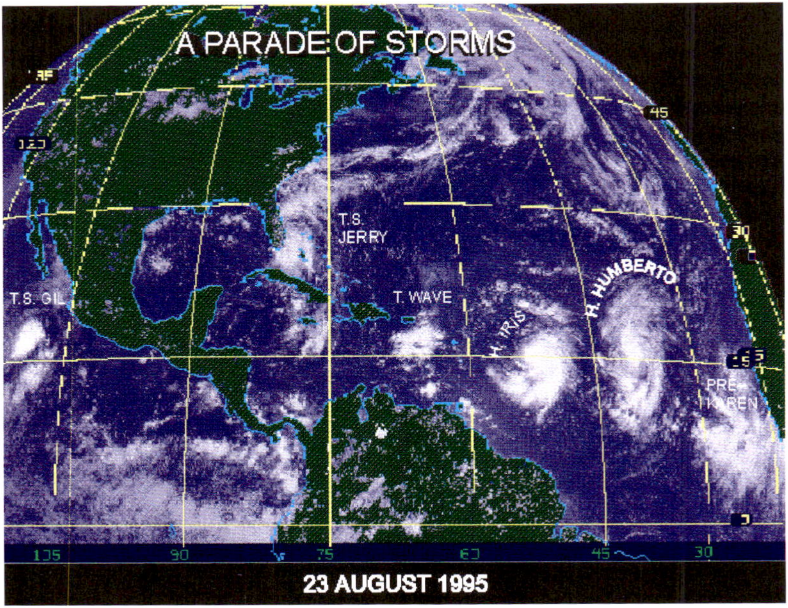

Figure 2.9 Harbinger of the future? A satellite image captures five tropical storm systems in various stages of development lined up as if awaiting a turn to make landfall in North America and the Caribbean *Source*: NOAA National Hurricane Center, Miami, Florida.

Heat waves

Stanley Changnon

"Before a potentially deadly heat wave arrives, we must have a reliable and accurate early warning system for emergency managers, the healthcare community, and the public". This statement from the director of the weather service for the southern United States underscores NOAA's ongoing interest in providing "life-saving services to the public".

- *Heat Outlooks:* Outlooks are issued for extreme heat conditions over a 48- to 72-hour period.

- *Heat Watches:* Watches are issued for a 24- to 48-hour period of extreme heat.

- *Heat Advisories:* Advisories (as well as "Excessive Heat Warnings") are issued for extreme heat conditions expected to occur within 24 hours.

Heat waves can be killers

Heat waves are commonly defined as periods of abnormally high temperatures (often associated with high humidity) that can last from a few days up to several weeks and can lead to mortality. They can also have an adverse impact on crops in the field. Heat waves producing deaths typically occur in areas where the heat event is infrequent and the local populace is not accustomed to its occurrence. The massive loss of life during the heat wave in the summer of 2003 in France, where few people were adequately prepared to cope with the unusually high temperatures and humidity encountered in a heat wave, reflects this reality.

The word "adjustment" is a key to understanding heat waves. For example, in 1995 the first heat wave since 1980 struck the American Midwest, leading to 718 deaths in five days. The same or more extreme conditions existed in the Deep South of the United States during that same year, yet only five heat-related deaths were reported there. The explanation for this discrepancy is that the unusually high temperatures and high humidity experienced in the Midwest in 1995 are common in the South. In fact, the critical threshold of apparent temperature for humans in the Midwest is 104°F (40°C) in the daytime and 86°F (30°C) at night, but these values are not considered critical in the southern United States, where the population is accustomed to and prepared for higher heat and humidity. In 1999, a similar heat wave struck the Midwest, resulting in only 258 deaths. The main difference between 1995 and 1999 was that many people in the Midwest had learned what to do and what not to do during heat waves. Furthermore, city and state institutions had also made adjustments and were better prepared for the heat.

During high-heat periods, four conditions affect the human body's ability to dissipate heat: air temperature, moisture in the air, air movement and radiation. Collectively, these four conditions create an "apparent temperature" that can be deadly to the unaccustomed or unprepared. The Heat Index (see fig. 2.10) developed by the US National Weather Service (NWS) provides a graphical representation of apparent temperatures as a guide to heat hazards.

In the United States heat waves cause an average of 1,000 deaths per year, many more than are caused by any other weather hazard. Deaths from heat waves are most common among the elderly. In Chicago, for example, the 1995 heat wave killed 525 people, 70 per cent of whom were 60 or older and 55 per cent of whom were 70 or older (fig. 2.11). A second key factor related to heat deaths is that most of those who die are residents of major urban areas. Again, in the 1995 heat wave 585 of the 718 people who died lived in the inner-city areas of either Chicago or Milwaukee. In fact, 80 per cent of all

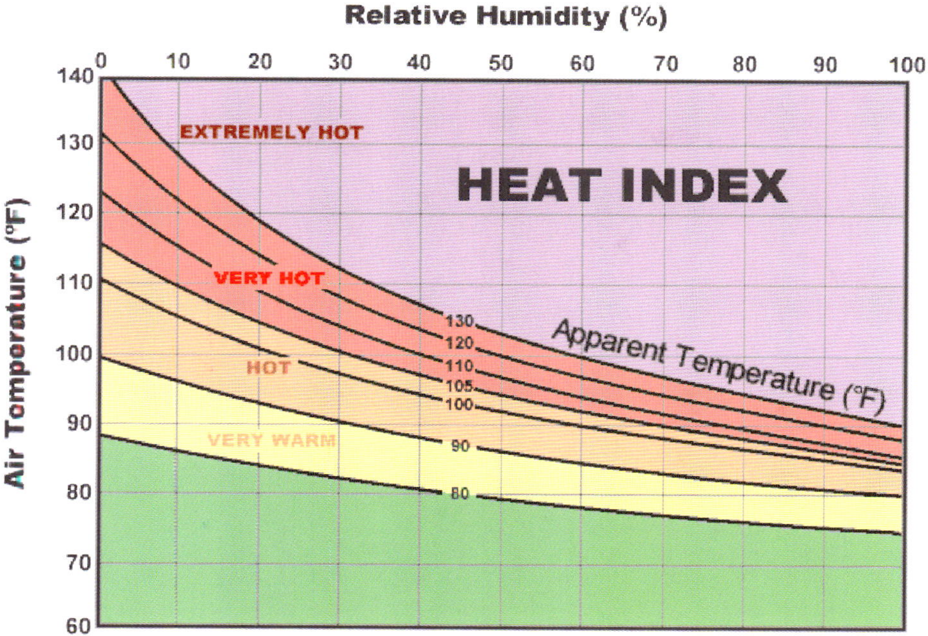

Figure 2.10 The Heat Index is an accurate measure of how hot it really feels when the effects of humidity are added to high temperature. Notably, the values in the index are only valid in the shade; up to 15°F (8°C) can be added for exposure to direct sunlight

Source: NWS.

heat-related deaths occur in large urban areas, where the number and density of structures and paved streets store daytime heat and slowly release it during the night in a process known as the "heat island effect". This effect intensifies during heat waves, causing night-time temperatures actually to increase by 2–3°C, which creates a third condition notable in heat waves – a lack of night-time cooling. One key to surviving high temperatures is cooling at night, which enables the body to rest and recover from the stress of an intensely hot day.

Figure 2.11 Cover of a book examining the 1995 heat wave disaster in Chicago, Illinois, US, that killed 525 people (Klinenberg, 2002)

Causes of heat-related deaths have also changed over time. Assessment of the extreme 1995 heat wave in Chicago included an examination of heat waves that occurred in 1931 and 1936. In these past events, apparent temperatures had actually been more severe than in 1995, but the loss of life was much less extreme. Several societal changes account for this discrepancy. First, in the modern world more people live to be 60 or older than in the 1930s, so there is a greater population of elderly people who are vulnerable to heat extremes. In addition, in the 1990s most of the elderly lived alone, whereas in the 1930s most lived with relatives who helped care for them.

A second societal change is related to nocturnal behaviour. In the heat waves of the 1930s, many poor people went to parks or the shores of Lake Michigan to sleep and cool off at night. In the 1990s, on the other hand, no one did this for fear of being harmed or robbed. For a similar reason, many people reported that in the more recent heat wave they were afraid to open windows at night to let cooler air into their residences because they feared the possibility of crime against themselves or their property. Finally, many of those killed by the 1995 heat wave were in low-income areas, and many poor elderly residents did not operate their air conditioners for fear that they would not be able to pay their electric bills.

Institutional adjustments to prepare for and respond to heat waves are critically important, and assessments of the Chicago and Milwaukee heat waves of 1995 further revealed that inadequate city warning and response systems were in place. For example, Chicago's declaration of a "heat emergency" occurred after the heat wave was almost over. Another problem that affected the cities was a power failures related to the high heat that resulted in a further loss of air conditioning as a palliative against the stifling conditions.

As suggested earlier, the sizeable decline in deaths in Chicago in the 1999 event compared to 1995 was largely attributed to improvements in the city's preparations and response systems. First, a special extreme weather committee was established. Additionally, the National Weather Service was assigned the role of predicting the development of dangerous conditions, and information advising local citizens of what to do and what not to do was published and widely distributed before and during the heat wave. The city also established numerous "cooling centres" in air-conditioned schools and city buildings where threatened people could go to sleep and rest. Finally, the city instituted a programme where city staff called and visited the elderly to alert them of the danger and learn if they needed help or transportation to a cooling centre.

In the end, surviving a heat wave requires two actions. First, individuals must adjust their lifestyles in response to apparent temperature extremes. Second, government agencies must develop effective warnings, education programmes and response activities.

European heat wave 2003

Janet Larsen

This case was originally published as "Setting the Record Straight: More than 52,000 Europeans Died from Heat in Summer 2003". (Larsen, 2006; reprinted with permission.)

Following a string of high-heat days and meteorologists' warnings that the summer of 2006 could be another scorcher, European public health officials and politicians revisited the devastating heat wave of 2003. The severely hot weather that withered crops, dried up rivers and fuelled fires that summer took a massive human toll. The full magnitude of this quiet catastrophe still remains largely an untold story, as data revealing the continent-wide scale have only slowly become available in the years since. All in all, more than 52,000 Europeans died from heat in the summer of 2003, making the heat wave one of the deadliest climate-related disasters in Western history.

Temperature records were broken in a number of countries in 2003 as Europe experienced its hottest weather in at least 500 years. The unusually warm weather began in June and culminated in an unrelenting wave of heat during the first two weeks of August (fig. 2.12). With both daytime and night-time temperatures remaining high, large numbers of vulnerable people, particularly the elderly, succumbed to the baking heat.

Hospitals were faced with unusually large burdens, and undertakers and funeral homes were overwhelmed. In France, doctors' warnings of a heat epidemic were largely quashed by the Ministry of Health's refusal to acknowledge the massive problem, reminiscent of

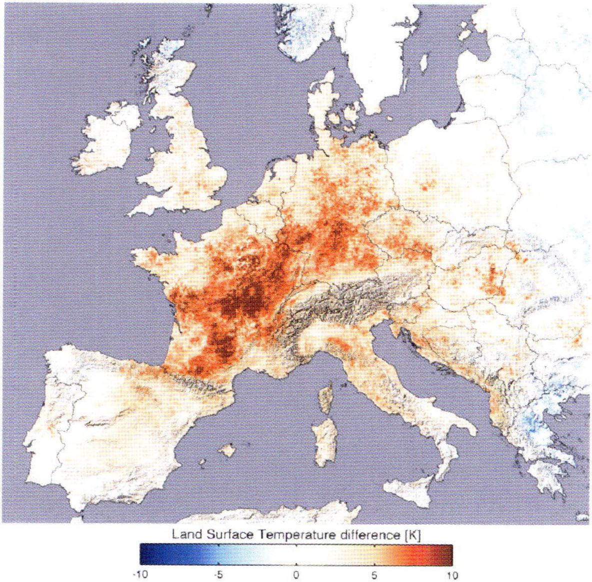

Figure 2.12 Difference in land surface temperature, calculated by subtracting the average of all cloud-free data during 2000, 2001, 2002 and 2004 from the averages measured in 2003, covering the date range 20 July–20 August

Source: R. Stocki and R. Simmon, NASA Earth Observatory.

the early political denial of the 1995 Chicago (United States) heat wave that killed more than 700 people in a matter of days. But as the bodies piled up, requiring makeshift morgues, "ignore and neglect" was no longer a viable option.

At the end of September 2003 the French National Institute of Health reported that in the first 20 days of August heat had killed more than 14,800 people. During the peak of the heat, fatality rates topped 2,000 a day.

Of the new information that has trickled out over the last few years, the biggest surprise has come from Italy. According to the Italian National Institute of Statistics, the summer of 2003 yielded more than 18,000 excess deaths when compared with 2002. In August alone, 9,700 fatalities were probably connected to the high temperatures, which in parts of the country averaged 9°C (16°F) warmer than in the previous year.

Altogether, new data boost Europe's heat-related mortality for the summer of 2003 by 17,000 over preliminary estimates (of 35,000) to a record 52,000 deaths.

Unlike hurricanes or tornadoes that leave obvious damage and death in their wake, not to mention vivid images for the media, heat waves are silent killers. Coroners' reports rarely list "heat" as the primary cause of death, even when high temperatures may have precipitated cardiovascular or respiratory failure or dehydration. Thus it is generally not until a heat wave is long over, when death counts can be compared to what would otherwise have been expected in a "normal" year, that we begin to learn the full human toll. Yet governments, reluctant to admit public health failures, often release such numbers with little fanfare.

While news reports gave estimates of a potentially large human death toll, it was not until well after the event that a more accurate tally became available. After facing criticism for its inadequate health facilities and lax government response, France became one of the first countries to release an epidemiological study revealing the true extent of the heat's damage.

Because reports of the heat wave's casualties trickled out of individual countries over more than two years following the actual event and never received widespread media coverage, however, policy-makers and the public at large have not grasped the full dimensions of the catastrophe and therefore underrate the risk of rising temperatures.

Indeed, after the 2003 event a number of European countries beefed up their heat-health alert systems and took additional measures to prepare for future heat waves. Following the public outrage over its 2003 failures, France's Ministry of Health announced increased funding for hospital beds and more jobs for health workers, as well as a renewed focus on care for elderly people who suffer the most during warm spells. The Spanish government's heat wave action plan includes an awareness campaign for social service and healthcare professionals, a voluntary register for people at high risk to receive special services and a daily mortality monitoring system.

Projections by the Intergovernmental Panel on Climate Change (IPCC), a global body comprising some 2,600 scientists, show more extreme weather events ahead as the planet heats up. By the end of the twenty-first century the world's average temperature is projected to increase by 1.4–5.8° Celsius (2.5–10.4° Fahrenheit). As the mercury climbs, more frequent and more severe heat waves are in store. Accordingly, the World Meteorological Organization (WMO) estimates that the number of heat-related fatalities could double in less than 20 years.

Early warning system for heat waves in Galicia, Spain

Lino Naranjo

A heat wave over Europe in the summer of 2003 brought an increase in disease and mortality, mainly among the most vulnerable populations – the elderly and the infirm. An estimated 52,000 casualties were directly related to the heat wave across Europe. France was the most affected country, with an increase in its mortality rate for this period of more than 130 per cent. In Spain, a total of 3,166 deaths, mainly in the northeast of the country, were related to the heat wave.

Galicia, in the northwest of Spain, was also hit by the heat wave, which mainly affected the population over 64 years old and especially those residents already suffering from circulatory problems, heart disease and respiratory dysfunction. As a consequence, the highest proportion of the 490 deaths in Galicia was in this demographic.

As a result of these severe impacts, several countries developed systems to monitor the consequences of heat waves in at-risk populations. Most of these systems proved to be more reactive than proactive, however, because they were based on mortality data from the 2003 event.

During the summer of 2004 Galicia developed its own multidisciplinary plan to mitigate the consequences of future heat waves. The plan has three main goals:

- to create an early warning system for heat waves,
- to define measures to cope with the impacts of heat waves in different risk sectors, and
- to ensure an effective response and cooperation among different levels of society.

An early warning system was created under the umbrella of the Galician environmental authorities, and Meteo-Galicia, the regional weather agency, assumed responsibility for developing the system operationally.

In the absence of a universally accepted definition of "heat wave", the EWS was designed to take into account Galicia's climate records, recognizing that one high-maximum-temperature event does not constitute a heat wave. Consequently, a heat wave alert is not released until three conditions are met: maximum temperatures exceed certain values (depending on locality); overnight temperatures are over 20°C; and these conditions persist for three days or more. This system is coupled with a graphic interface of the output of the operational high-resolution numerical model, ARPS, that runs twice a day at Meteo-Galicia.

A colour scale for different risk levels was also defined based on the METEOALARM warning schedule (see fig. 2.13) of the Network of European Meteorological Services (Eumenet), which is widely used in the European Community.

In Galicia's case, colour levels were modified to improve the details of the warning.

- *Green level*: Normal conditions.

- *Blue level* (pre-exposure): Observed conditions are normal but the model predicts that the following three days will be over established risk levels.

- *Yellow level* (exposure): Observed conditions and predictions for up to three days are over established risk levels.

- *Orange level*: Currently observed conditions are in either the yellow or red level; however, the model predicts a return to normal levels in a day or two.

METEOALARM warning schedule
Redagno 2006
Folie 17

	Damage / Impact	What to do?	Used how often? (per approx. 300 kkm^2)	Meteo treshholds **Rain** area related + damage related
Green	- - -	usual phenomena		
yellow	exposed objects (avoidable)	caution with exposed activities	> 30 per year	> 54 mm/12h
orange	general damages (not avoidable)	keep informed in detail, follow advice of authorities	1 to 30 per year	> 80 mm/12h
red	extreme damage and /or casualties *extreme damage (mostly) on large areas, threatning life and properties* (not avoidable, even in otherwise safe places)	**follow order of authorities under all circumstances** **be prepared for extraordinary measures**	less then 1 year *for large (5000km^2) scale phenomena*	> 140 mm/12h

Figure 2.13 METEOALARM warning schedule

- *Red level* (critical): Previous and current conditions are over the established risk level and will remain so for at least three days according to the forecast model.

Output from Galicia's EWS for heat waves is published using a public webpage that graphically maps the risk levels for different municipalities (councils). This map is continually revised by the operational forecaster using a graphic interface to correct possible gaps caused by bugs in or erroneous output from the model.

Warnings in Galicia are linked to an action plan for coping with the effects of a heat wave. This plan involves several departments in the Galician autonomous government, which have created a joint committee to monitor situations and take appropriate measures that provide advance information about locations affected and populations at risk, and anticipate increased demands for social assistance and healthcare at medical facilities. When a warning is released, the action plan is actuated, prompting the Civil Defence and networks of medical units, social assistance organizations and volunteer groups to monitor at-risk populations. Such actions include disseminating information and broadcasting advice through the media and local health centres.

In the summer of 2006 Galicia's heat wave EWS proved effective when the potentially severe impacts of observed heat wave conditions were mitigated by the issuance of 18 warnings at several colour levels over a period of 27 days. Fewer than 60 people were negatively affected by the extreme heat events.

Tornadoes

Margaret Lemone

- *Tornado Outlook:* Outlooks are issued when there is a high potential for severe thunderstorm activity capable of producing tornadoes.

- *Tornado Awareness:* The first stage of an early warning of possible tornado activity. Awareness notification is issued.

- *Tornado Watch:* Watches may cover an area of 51,800–103,600 square kilometres (20,000–40,000 square miles) and last four to six hours. They imply that conditions for severe weather are favourable but not guaranteed, and can be cancelled, revised or reissued. Watch notification is issued.

- *Tornado Warning*: Warnings mean that a tornado has been sighted or indicated by weather radar. Severe weather is imminent.

Tornadoes produce the most violent winds in nature. According to the *Glossary of Meteorology*, a tornado is a "violently rotating column of air in contact with the ground". Usually, but not always, such a column is visible as a funnel-shaped cloud hanging from the base of a thunderstorm (cumulonimbus) cloud. The visible funnel can be made up of cloud drops, dust and debris carried upward by the converging winds that form the tornado. Looking from the top down, the winds in a tornado normally rotate counter-clockwise in North America, although rotation in the opposite direction has been observed.

Although a typical tornado is only about 0.4 km (0.25 mile) in diameter and remains on the ground for less than five miles, they can reach over 1.62 km (1 mile) in diameter and remain on the ground for much longer distances. Winds in a tornado can reach up to 480 k/h (300 mph), though typical wind speeds are less than 240 k/h (150 mph).

Interestingly, actual tornado wind speeds can only be approximated until a detailed engineering analysis of damage to structures has been completed in the weeks following an event. Because tornadoes of different wind speeds have been known to cause similar damage to structures and vegetation depending on various factors independent of the event itself (e.g. building materials, vegetation root systems, etc.), only through such detailed analysis can the actual intensity of the event be determined. In any case, the most intense tornadoes have been known to lift houses from their foundations, carry cars several hundred feet into the air and tear the bark off trees.

Figure 2.14 A large tornado touched down just south of Dimmitt, Texas, US, on 2 June 1995
Source: Harald Richter/NOAA Photo Library.

Tornado damage is measured by the Fujita Tornado Damage Scale, normally referred to as the F-scale. Damage categories in this scale are:

- light (F0)
- moderate (F1)
- considerable (F2)
- severe (F3)
- devastating (F4)
- incredible (F5).

An enhanced F-scale developed by a team of meteorologists and wind engineers and consisting of 28 detailed categories was implemented in February 2007.

On average, more than 1,200 tornadoes resulting in 70 deaths, 1,500 injuries and over $400 million in damages occur in the United States every year. They can happen at any time of year, at any time of day and in any part of the world, but they occur most frequently in the central United States in late afternoon in the spring and summer. Notably, the geographic range of maximum tornado threat moves northward with the change of season typically occurring in spring in the southern United States and in summer in the northern United States. The most favourable conditions for the formation of tornadic thunderstorms are a significant change in wind direction, speed and altitude; warm, moist air in the lowest few kilometres of the atmosphere; and the development of a trigger or catalyst, such as a cold front.

People can do little to protect structures from the fury of a tornado. With adequate warning, however, they can avoid injury – which mainly results from flying debris – by moving to a protected area. In a building, a relatively safe location is in the basement, an inner room or a hallway on the lowest floor and away from windows. If a tornado forms while you are driving, the safest place is out of your vehicle and in a ditch alongside the roadway. Do not seek shelter beneath a highway underpass. By their nature, mobile home communities are extremely vulnerable to high winds, so these homes should be abandoned for the protection of the nearest storm shelter when severe weather approaches.

In the last 10 years, tornado warning times have doubled thanks to a nationwide network of radars (WSR-88D) and an increase in the sophistication of their use. Effective, prearranged communication channels through radio and television broadcasts, as well as warnings by local and state police, have also contributed to increased warning times. Successful public education and outreach campaigns by the US National Weather Service, television and radio personalities, emergency preparedness officials and school teachers have proven invaluable in educating people about what to do at home, in school and in the workplace when a tornado alert is issued. Studies show that these lessons are taken quite seriously in areas where people have first-hand experience of tornadoes.

Although the United States is beset by a major share of global tornadoes annually, tornadoes also occur worldwide in such locations as Uruguay, Argentina, Southern Africa, eastern Australia, New Zealand, coastal China, Japan, Bangladesh and Europe.

El Niño Southern Oscillation

Michael McPhaden

Several monitoring schemes and modelling groups are constantly seeking to forecast changes in sea surface temperatures in the central and eastern equatorial Pacific to provide insight into the possible development of an El Niño. Unlike most other hydrometeorological early warning systems, forecasts of El Niño onsets do not use terms like advisories, watches and warnings in a systematic or hierarchical scale progressing from a possible (foreseeable) episode to an imminent event. Different groups, however, do mention either an El Niño advisory, an El Niño forecast, an El Niño watch or an El Niño update.

The most important known factors in the forecasting of an El Niño episode are the monitoring of sea surface temperatures (fig. 2.15), surface winds, upper ocean heat content and sea-level pressure changes in the tropical Pacific Ocean along the equator. Once the scientists who monitor these changes have identified the onset of an episode, reports tracking its development along with projections about its future course of action and impacts are issued on a weekly basis.

The magnitudes of El Niños, which have been classified as weak, moderate, strong and very strong, are determined by groups that use different criteria: the area encompassed by an anomalous sea surface temperature warming, the amount of damage an event causes worldwide, and so forth. Whatever the criteria, projections of magnitude are always highly speculative at the onset of an El Niño episode, becoming more definitive as the event develops.

Perhaps even more important than forecasting the likelihood or magnitude of an El Niño many months in advance of its onset, however, is forecasting its regional and local impacts in areas known to be affected by such episodes. In fact, identification of the onset of an El Niño raises worldwide awareness of a number of teleconnected (correlated) events and triggers a veritable cascade of early warning systems for a range of impacts, including droughts, floods, fires, infectious disease outbreaks, fish catches, frosts, food shortages and even famines.

El Niño: What it is

Every two to seven years, the westward winds along the equator weaken and sometimes reverse, blowing eastward. As a result, the warm water that has "piled up" in the western

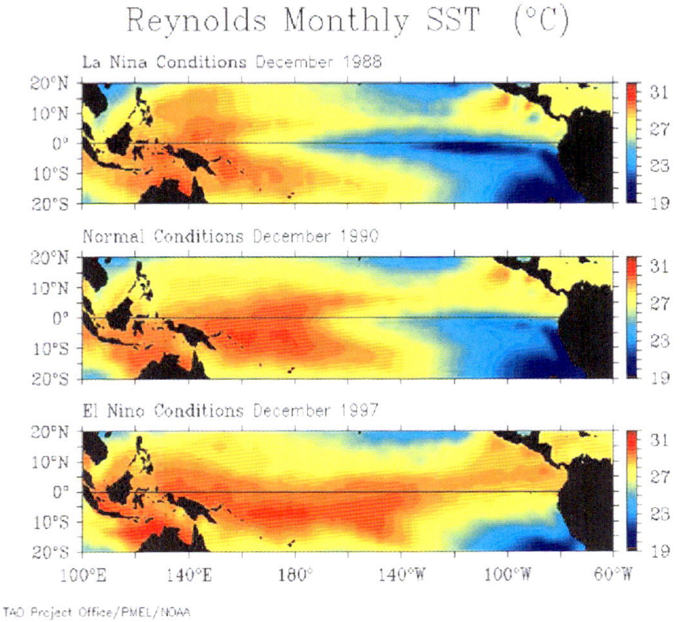

Figure 2.15 A real-time global sea surface temperature (SST) analysis developed by Richard Reynolds of the US National Climatic Data Center indicates normal ocean temperature conditions (middle), conditions during the 1988 La Niña (top) and conditions during the 1997 El Niño (bottom)
Source: NOAA.

part of the basin begins to shift towards the central and eastern Pacific. The sea level drops in the west and rises in the east, reducing the upwelling of water from below the surface east of the international dateline that normally cools sea surface temperatures (SSTs). An El Niño occurs at this point, when SSTs in the central and eastern Pacific increase by 2–3° Celsius or higher along the Peruvian and Ecuadorian coasts. In 1997–1998, during the strongest El Niño of the twentieth century, SSTs along this stretch of the Pacific increased by 5–6° Celsius.

Today, the term "El Niño" is used to describe both the localized warming of the coastal ocean off Peru and Ecuador and the much larger, basin-wide air-sea interactions that occur across the equatorial Pacific Ocean (fig. 2.16). It represents the warm phase of the so-called El Niño Southern Oscillation (ENSO) cycle. The cold phase of ENSO is known as La Niña. The Southern Oscillation refers to a seesaw in atmospheric surface pressure between the eastern Pacific and the western Pacific/eastern Indian Ocean region. A convenient index for ENSO, which provides a measure of trade-wind intensity, is the pressure difference between Tahiti (French Polynesia) and Darwin (Australia). When sea-level pressure is high in Tahiti it is usually low in Darwin, and vice versa. Australians tend to use the Southern Oscillation index as an early warning of drought, whereas Peruvians tend to rely on changes in sea surface temperatures in the eastern equatorial Pacific to prepare for flooding in northern Peru.

El Niño: What it does

The intensity of an El Niño event can vary from weak to very strong. Strong events are more likely to influence climatic conditions far from the Pacific Basin, whereas weak ones are likely to have their strongest impacts in Pacific Rim countries.

El Niño conditions typically bring devastatingly heavy rains to Peru's normally arid coastal areas and droughts to Bolivia and southern Peru. They have also been associated with

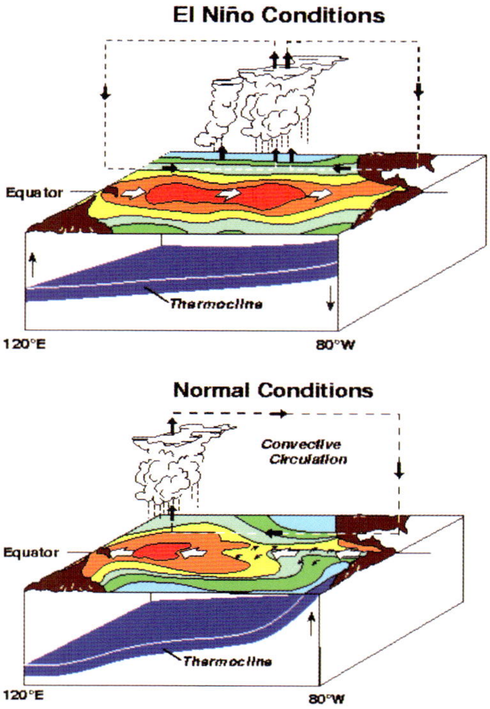

Figure 2.16 Conditions leading to an El Niño

severe drought in the semi-arid Brazilian *Nordeste* and flooding in the southern regions of the country. El Niño has usually, but not always, also been associated with droughts in Australia, Indonesia, the Philippines, Papua New Guinea, Southern and Eastern Africa and the Horn of Africa. El Niño-related flooding often occurs in Argentina, Paraguay and Uruguay.

The 1997–1998 El Niño episode was linked to major forest fires in Indonesia and a resultant haze across Southeast Asia. These fires encompassed about 9 million hectares (22 million acres) and, although blamed on El Niño-related drought conditions, were later discovered to have been intentionally orchestrated by unscrupulous businessmen (and government ministers) who paid people to set the fires, knowing that a forecast of El Niño would translate into "no rain" to extinguish them. These businessmen and ministers had planned to buy the burned-out forest areas for their own purposes and gain.

The worldwide costs attributed to the climate and weather anomalies ostensibly spawned by the 1997–1998 El Niño are estimated at between US$32 billion and $96 billion.

A list of likely "El Niño hotspots" includes the following.

- Drought in Zimbabwe, Mozambique and South Africa.
- Warm winter weather in the northern half of the United States and southern Canada.
- Heavy rains in southern Ecuador and northern Peru.
- Drought in northeastern (*Nordeste*) Brazil.
- Flooding in southern Brazil.

- Drought and fires in Indonesia.
- Drought in the Philippines.
- Drought in various South Pacific island nations.
- Heavy rains in the island states of the central equatorial Pacific.
- Drought in eastern Australia.
- Heavy rains in southern California.
- Fewer than average hurricanes in the tropical Atlantic.
- Disruptions in Pacific marine ecosystems and fisheries.
- Coral bleaching worldwide.

Vector-borne disease

Mary Hayden

Infectious vector-borne diseases are a major cause of mortality in the developing world, with malaria accounting for 1.2 million deaths annually, mostly among African children under the age of five. Other arboviruses such as dengue fever and Rift Valley fever (RVF) are of global concern because of associated morbidity in addition to mortality. An estimated 50–100 million cases of dengue are reported annually, of which 500,000 are the more severe manifestation, dengue haemorrhagic fever (DHF). After several years of decline, RVF has once again emerged in East Africa in the aftermath of extensive flooding in 1997–1998 and 2006–2007. Concern exists that these mosquito-borne diseases will continue to proliferate in the tropics and subtropics, and expand their geographic range into the temperate zones in response to extreme weather episodes, climate variability, global warming and reduced spending on public health.

A new paradigm for reducing the impact of vector-borne diseases is long overdue, and newly emerging transdisciplinary efforts integrating ecological, biological and social factors offer promise in this area. Specific issues currently being addressed are local ecology (both human and vector); the changing nature of vector behaviour over time in response to evolving social and ecological systems; and the resultant implications for policy. Recognition of the necessity for successful community-based field research that addresses the need for continued programmatic vector surveillance and multidirectional feedback to guide policy is critical to ensuring that an ecosystems approach is systematically implemented. The role of governments in coordinating efforts aimed at Integrated Vector Management (IVM) cannot be underestimated, as both directed "top down" and grassroots "bottom up" approaches are needed to develop and sustain programmes.

Early warning systems are being developed to monitor climate variables, such as monthly rainfall and temperature, related to mosquito-borne disease transmission. These climate data are being developed into seasonal risk maps which could potentially reduce vulnerability when coupled with mosquito surveillance, field epidemiology focused on understanding social and ecological factors responsible for vector abundance and human-vector contact, and public educational outreach.

Development of these new EWSs is critical, as, for example, a warning system may have reduced recent transmissions of RVF in East Africa. An effective EWS may have alleviated much of the misery because the virus not only affects humans but also livestock, leading to heavy losses that increased the suffering caused by the outbreak. Furthermore, in both 1997–1998 and 2006–2007 floods were a precursor to disease onset, so better monitoring and detection of the development of the disease in livestock in the aftermath of extreme weather could have provided an early warning to prevent transmission to humans. In this way, a climate early warning system coupled with an integrated approach to disrupting transmission through health education regarding the slaughter of animals, heightened mosquito abatement and, in the case of RVF, access to animal vaccines has excellent potential for reducing disease impacts.

Climate-based malaria early warning system in Ethiopia

Tsegay Wolde-Georgis

Malaria annually affects hundreds of millions of people worldwide. The vast majority of deaths from malarial infection occur in sub-Saharan Africa (SSA), where malaria also affects socio-economic development through lost production activities. Ethiopia, where about 10 million people are affected each year, is one of the malaria epidemic-prone countries in Africa (fig. 2.17). The disease has a significant impact on the country's development, representing about 20 per cent of yearly medical cases, putting pressure on the health services and leading to labour shortages during the harvest.

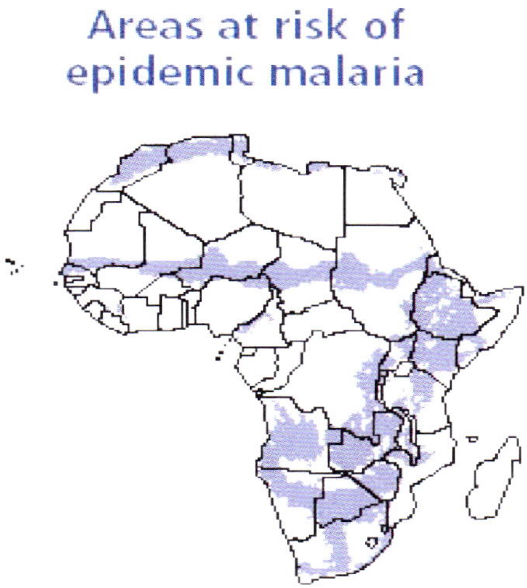

Figure 2.17 Areas at risk of epidemic malaria
Source: FAO.

Ethiopia's health strategy focuses on prevention, and forestalling malarial epidemics is a core activity. When malaria epidemics are foreseen, local authorities and non-governmental organizations (NGOs) advise communities to clear standing water to deny mosquitoes their breeding areas. In addition, authorities have begun to distribute insecticide-treated bed-nets (fig. 2.18) widely, and, beginning in 2006, to use the latest anti-malarial medicine, Coartem, which has proven 99 per cent effective in curing malarial infections.

Figure 2.18 A girl in Karo Duss village, Ethiopia, under an insecticide-treated mosquito net
Source: UNICEF.

According to the Ethiopian Ministry of Health, in 2007–2008 alone 20.4 million treated mosquito nets were distributed to 10 million households and 4.6 million doses of Coartem were administered. Significantly, many health centres now also use cheap rapid diagnostic test (RDT) kits to identify malaria cases on the spot and begin immediate treatment regimes.

Because of the government's commitment to prevention and its important partnerships with key global funders, Ethiopia now has a chance to eradicate malaria as its number-one killer. As a result, Ethiopia has been admitted as a member of the board of the Rollback Malaria Partnership, an initiative of the UN World Health Organization (WHO).

The development of the malaria early warning system

Until recently, climate information had never been used to forecast malaria epidemics in Ethiopia, even though such epidemics have long been known to be climate-sensitive. Beginning in 2006, however, the Ethiopian Ministry of Health (MoH) mandated that the National Meteorological Agency (NMA) collaborate on the development of a malaria early warning system. The objective was to enable the MoH to disseminate malaria early warnings to at-risk populations and decision-makers.

The collaboration between the MoH and the NMA originated from previous research funded by the WHO and Harvard University that identified the relationship between meteorological variables and malaria epidemics in Ethiopia. This research investigated the hypothesis that anomalies in precipitation, temperature and relative humidity are potential triggers for malaria transmission in the highland dry areas of the country. The research concluded that climate variables provide the strongest correlation for malaria transmission, identifying a lag time of two months between the end of the main rainy season (*kiremt*) and the occurrence of an outbreak. Identification of this lag provided an opportunity to develop a malaria warning system based on local meteorological factors that, according to the study, would prove cheap, practical and effective.

Because the basis of the early warning system is the supposition that malaria outbreaks typically occur "when the monthly accumulation of precipitation (rainfall) is at least 80 mm, the monthly mean temperature is between 18°C and 32°C, and the monthly relative humidity is at least 60%" (www.malariajournal.com/content/5/1/38), the MoH/NMA collaboration focused on strengthening meteorological data collection and analysis. The MoH allocated funding of US$500,000 in 2006 specifically to upgrade 319 meteorological stations that only measured rainfall (class IV) to stations that could also monitor minimum and maximum temperatures (class III). Fourteen sentinel stations for early detection in malaria-prone districts were also established (www.wmo.ch/pages/prog/amp/pwsp/documents/Connor.pdf). These stations are notable because, traditionally, meteorological stations in Ethiopia had been located in big towns and populous areas, but the stations identified for upgrade by the MoH were chosen primarily for their locations' propensity for malaria outbreaks.

To analyse the meteorological data, the NMA has divided the country based on precipitation, humidity and temperature conditions, continually developing maps and providing colour images of areas where these three variables overlap, indicating a potential outbreak hotspot. Daddi Jima of the Malaria and Other Vector-borne Diseases Control Unit at the MoH observed that climate analysis from the NMA provides better information than previously available on malaria outbreaks. More data are needed, however, on such variables as topography, water and other factors like population mobility.

This was the first time an Ethiopian ministry had demanded the services of the NMA by funding its contribution, and the MoH was not initially pleased with the product it received from the NMA, which was vague, generalized and not useful for adequate intervention planning. The ministry demanded more refined climate data that would be more useful for its activities. According to Tesfaye Gissila, head of the Developmental Meteorology Department at the NMA, his institution sent experts to the MoH to discuss and assess what climate information would be most relevant for malaria early warning, which was a challenge for the NMA since it was being asked to provide a highly specific dataset relevant to the decision-making activities of the MoH.

Following additional meetings between the malaria control experts at the MoH and climate experts at the NMA, an agreement was finally reached in January 2007 specifying that the NMA would provide detailed and continuous monthly climate information for the *Health Sector Bulletin*. The bulletin is distributed to the MoH, which, in turn, disseminates it to regional malaria control departments with a cover letter and directives. The information is also posted on the Internet through RANET and the MoH and NMA websites.

The climate information in the bulletin does not necessarily identify the timing of the outbreak of malaria, since non-climate interventions such as the timely delivery of mosquito nets and medicine can prevent outbreaks; however, the early warning bulletin from the

NMA does indicate the potential for malaria epidemics, which the MoH uses for intervention planning purposes. Its usefulness is expected to increase in the coming years as it is tailored to meet the needs of specific local decision-makers and populations more accurately.

Figure 2.19 presents a sample risk map combining precipitation, humidity and temperature conditions for September 2007. The areas overlain with square patterns and with a yellowish background satisfy the criteria established for climatic conditions favourable for malaria epidemics, triggering action by the Ministry of Health.

According to the Ministry of Health, this information will lead to the following actions:

- declaration of a health emergency
- execution of a plan to purchase drugs and mobilize treatment to hotspots
- increased prevalence of malaria surveillance through weekly surveys
- heightened distribution of insecticide-treated mosquito nets
- broadcasts of education campaigns about malaria prevention.

Experts at both the MoH and the NMA believe that more must be done to streamline the malaria early warning system. They cite the lack of skilled human resources, such as biometeorologists, in the "boundary" areas to collect data, and the need for coordination with the Ministry of Agriculture, whose extension agents frequently interact with farmers who could provide significant data. Other constraints cited by experts from the two institutions include lack of access to mapping software as well as "brain drain" from the country. Even

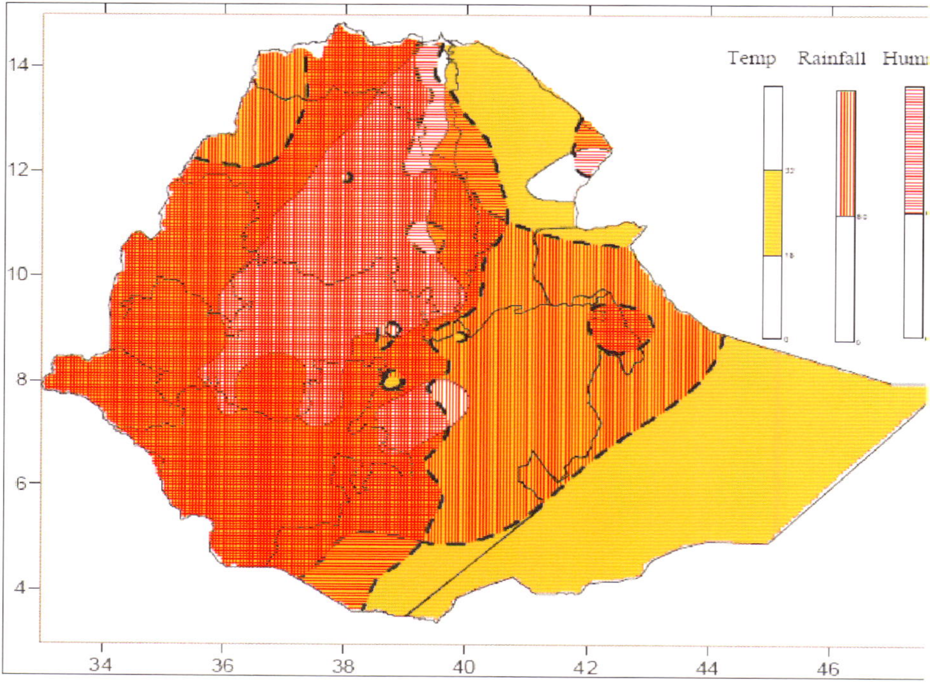

Figure 2.19 Climate information for the health sector
Source: National Meteorological Agency (2007).

though there has not yet been an evaluation of the effectiveness of Ethiopia's malaria early warning experiment in epidemics, the MoH's demand for climate information to support its activities is a bold and promising initiative.

Avian influenza

Douglas Pattie

Early warning of outbreaks and the capacity for prediction of spread to new areas are an essential prerequisite for the effective containment and control of epidemic animal diseases, including zoonoses, which are diseases that can pass between animals and humans. All over the world, weak disease surveillance systems and the inability to control major diseases at their sources have contributed to the spread across geographical boundaries of diseases confined to livestock, such as avian influenza (AI).

Highly pathogenic avian influenza (HPAI) H5N1 was first recognized as an important animal disease in August 1997 following an outbreak in Hong Kong when 18 people became infected, with six cases proving fatal. In order to halt the outbreak, a mass cull of all poultry in Hong Kong was carried out, a process that was repeated during another outbreak in May 2001. Since then, the Hong Kong government has taken a range of steps to reduce the risk of further AI outbreaks. The economic losses associated with HPAI across parts of Asia, Europe and Africa are estimated in billions of dollars. More disturbingly, as the virus moves between wild birds, domesticated poultry and people, the potential intensifies for a pandemic like the 1918 influenza epidemic that killed an estimated 40 million people worldwide.

Cataloguing the events in relation to occurrences of HPAI in poultry, wild birds and humans (on a global scale) reveals that its incidence has steeply increased since the beginning of 2004. Wild-bird case frequency peaked in early 2006 – the year following a mass die-off at Qinghai – with a strong drop in frequency since. Very few cases were seen in 2007. The websites of the World Health Organization (WHO) and the UN Food and Agriculture Organization (FAO) provide good overviews of these events, both in listings and in map format.

The outbreak of H5N1 was first identified in Indonesia in 2003 through a series of diagnoses from clinical symptoms, pathological changes and laboratory examinations. Records indicate that from July 2005 to March 2006 there were 30 cases of HPAI infecting humans, of which 22 resulted in death. Between May 2005 and May 2007, 5,432 cases of HPAI H5N1 in wild birds were reported to the World Organization for Animal Health (OIE). Of these, only 2,198 (40 per cent) were identified at the species level, 2,650 (49 per cent) were assigned to groups such as wild ducks or gulls and 584 (11 per cent) were described as infecting migratory or wild birds. In all, the virus has infected wild birds or poultry in more than 60 countries in Asia, Europe and Africa. Twenty-one of these countries reported outbreaks in 2007. The continuing spread of H5N1 appears related to the legal and illegal movements of poultry and their products as well as to the movements of wild birds (Liu et al., 2005).

Lack of hygiene, overstocking and mixing of different domestic animals greatly increase the risk of spreading the infection. Movements of people (e.g. farmers, veterinarians and even journalists and tourists) and legal and illegal trade in caged birds are also factors in the spread. Clearly, trade in domestic poultry has been a crucial factor, even for the transmission of HPAI over long distances and across continents; however, numerous species of wild birds, especially waterbirds, have also proven to be susceptible to infection by H5N1 of the Asian lineage. Close contact between wild birds and poultry can lead to cross-infection, both from poultry to wild birds and from wild birds to poultry. Significantly, the continued loss of wetlands around the globe may be forcing many wild birds to alternative sites, like farm ponds and paddy fields, bringing them into direct contact with chickens, ducks, geese and other domestic fowl. Additionally, species that live in and around poultry farms and human habitations may serve as *bridge species* that could potentially transmit the virus between poultry and wild birds; notably, however, recent samplings of 5,000 wild waterbirds in African wetlands support the view, since no evidence of HPAI was found, that wild birds probably play a relatively minor role in the spread of avian influenza. This view is consistent with the fact that the northward migration of wild birds from Africa to Europe in the northern spring of 2006 did not cause any major outbreaks.

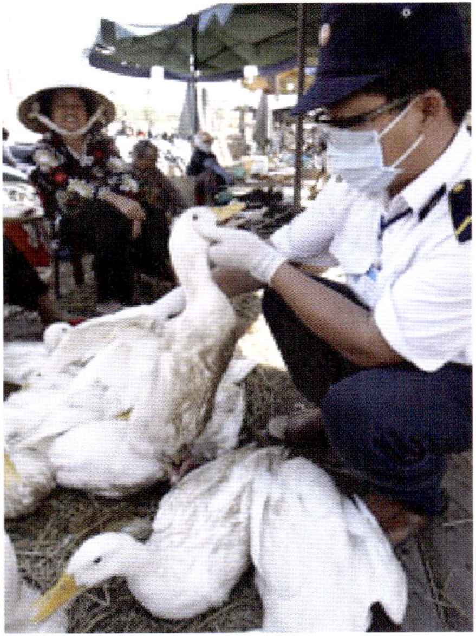

Figure 2.20 A veterinary officer inspects live ducks in a market in Viet Nam
Source: Hoang Dinh Nam/FAO.

Developing wildlife early warning and surveillance programmes and enhancing biosecurity in relation to avian influenza raises preparedness issues common to risks from other zoonotic diseases such as Japanese encephalitis, West Nile virus, Crimean-Congo haemorrhagic fever and equine encephalitis (Venezuelan, Eastern or Western). Poor identification, risk assessment and reporting to the OIE remain major concerns. Analysis of recent reports to the OIE where wildlife are part of the outbreak or die-off records shows that these reports often lack species identification and information on the precise location and timing of infection or the means by which cases are detected. These deficiencies constrain improved understanding of the H5N1 HPAI epidemiology. Worryingly, while most attention has been focused on HPAI H5N1, other H5 and H7 subtypes, both of which are avian-borne diseases, also pose major risks.

The creation of more strategic approaches to developing integrated HPAI early warning systems at regional or wider scales is being encouraged through monitoring appropriate parameters, including migratory patterns of higher-risk species and the risk of such species mixing either with other wild species and/or with poultry. Openly accessible data and information on the location and extent of avian influenza surveillance and results in wild birds are important to help international understanding of the ecology of this virus. One model for this is the Global Avian Influenza Network for Surveillance (GAINS), an open database and mapping system that presents a desirable wild-bird module for the FAO/OIE/WHO-led Global Early Warning System (GLEWS) for transboundary animal diseases, including zoonotics. GLEWS is based on the concept that dealing with a disease epidemic in its early stages is easier and more economical than having to deal with it once it is widespread. From a public health perspective, early warning of outbreaks with a known zoonotic potential will enable control measures that can prevent human morbidity and mortality. For zoonotic events, alerts of animal outbreaks can provide direct early warning so human surveillance can be enhanced and preventive action taken. Importantly, however, circumstances may arise where human surveillance proves more sensitive and alerts of human cases precede known animal incidences of a disease.

A number of early warning field projects are examining, through the use of satellite telemetry and remote sensing information, migratory patterns of key wildlife species thought to play a role in the spread of HPAI H5N1 virus. Since 2006 an integrated development of global early warning strategic surveillance programmes based on risk assessments has taken place throughout the European Union. This programme is planned for development under the framework of the EU-funded New-Flubird project. The ornithological component of the project, under the lead of Wetlands International, will identify higher-risk species, describe their migrations and identify strategic sites for monitoring. A reporting system is being established that will gather the information from these sites and allow integration with other initiatives, including the GAINS system.

Severe weather

Rebecca Morss

- *Convective Outlooks:* The US National Weather Service (NWS) Storm Prediction Center (SPC) issues day 1, day 2 and day 3 outlooks indicating where in the United States severe thunderstorms may develop in the next six to 72 hours.

- *Severe Thunderstorm Watch:* When severe thunderstorms – storms with large hailstones and damaging winds – are possible in the next few hours, the SPC issues a watch, which generally covers 51,800–103,600 square kilometres (20,000–40,000 square miles) and lasts three to six hours. Watches are usually issued where severe weather may be significant or when the severe weather threat is expected to last for several hours; watches may not be issued for isolated severe weather. In rare situations, when particularly intense damaging winds are possible, the watch may include the wording "this is a particularly dangerous situation". Severe thunderstorm watches notify emergency management personnel and the public of the risk of severe weather, so they can take appropriate action or monitor the situation as it evolves over several hours.

- *Severe Thunderstorm Warning:* Issued by local NWS offices when severe thunderstorms are imminent or already occurring in a specific area, severe thunderstorm warnings notify people in affected areas that they should seek immediate and appropriate shelter. Tornadoes can occur under severe

Figure 2.21 Often associated with violent weather, a supercell formed over Miami, Texas, US, on 19 June 1980
Source: NOAA Photo Library.

thunderstorm watches and warnings, but tornado watches and warnings are issued separately when conditions are especially favourable for tornadoes.

When conditions appear favourable for severe weather development, the SPC also issues a *Mesoscale Discussion* (MCD) that describes what is currently happening, what is expected to happen over the next few hours and the meteorological reasoning for the forecast. For severe thunderstorms, MCDs are generally issued one to three hours prior to a watch. MCDs are, however, also issued for hazardous mesoscale aspects of winter weather, such as heavy snow, blizzards, freezing rain and occasionally heavy rain. Winter weather and heavy rain MCDs are intended to provide additional meteorological guidance through a potentially hazardous situation to local NWS offices and other interested people; local NWS forecasters are, however, charged with providing detailed local forecasts and warnings as needed (Storm Prediction Center, 2006).

In general, the term "severe weather" describes any type of weather that is potentially hazardous (i.e. that threatens life or property). Examples include heavy rain or snow, freezing rain, high winds, excessive heat or cold, hazardous fire weather and thunderstorms. Severe weather can occur at any time throughout the year, usually in conjunction with extra-tropical large storm systems, hurricanes, tropical cyclones or warm-season conditions favourable for thunderstorms.

From a societal perspective, however, severe weather most frequently refers to potentially hazardous weather associated with severe thunderstorms, including hail, heavy rain, lightning, tornadoes and high winds. The NWS defines a *severe thunderstorm* as any storm that produces one or more of the following:

- a tornado
- winds of 50 kts (92.6 km/hr or 57.5 mph) or greater
- hail 0.75 inches (19 mm) or larger in diameter.

Although lightning is also hazardous and often occurs during severe thunderstorms, it is not included in the NWS criteria because "ordinary" (non-severe) thunderstorms can also produce significant lightning. Similarly, flood-inducing rainfall may also occur during severe thunderstorms, but is not included in the NWS criteria for the same reason. In contrast, tornadoes, damaging winds due to downbursts and damaging hail occur almost exclusively in the context of severe thunderstorms.

Thunderstorms require three ingredients to develop.

- *Moisture.* Water vapour in the lower levels of the atmosphere that can form clouds and rain.
- *Unstable air.* Air at lower levels of the atmosphere that has the capacity to rise rapidly because the air above it is relatively cold and dry.
- *A lift mechanism.* Something to lift the moist, unstable air off the surface, such as a cold or warm front, variations in terrain, variations in heating at the surface or remnants of previous thunderstorms.

A typical thunderstorm is about 24.3 km (15.1 miles) in diameter and lasts an average of 30 minutes. These "single-cell" storms are generally not strong enough to produce severe weather; however, single thunderstorm cells can organize to form *multicell storms* that contain clusters of cells in various stages of thunderstorm development. Such storms can last for several hours, producing large hailstones, flash flooding, damaging winds and isolated tornadoes. Thunderstorms can also organize in a line, called a squall line, which can persist for many hours as new cells continually form (fig. 2.22).

The most damaging severe weather tends to be associated with *supercell thunderstorms*, a special kind of storm in the northern hemisphere that has a counterclockwise rotating centre and can last for many hours. Along with heavy rains and extreme downdraught

Figure 2.22 An intense line of thunderstorms moved across the Middle Atlantic on 15 May 2004. Strong straight-line winds occurred in advance of the storms
Source: Photo by Kevin Ambrose.

winds, supercells cause nearly all of the large hailstones (golf-ball size) and significant tornadoes in the United States. These major events generally form when the atmospheric environment contains a fourth ingredient.

- *Strong wind shear.* Winds that significantly change direction or increase in speed with height (fig. 2.23).

Hail forms when strong updraughts (rising air) in severe thunderstorms carry raindrops up near to the top of the storm where air can be cold enough for water to freeze. Large hailstones form when the updraught is strong and sustained enough to lift the hailstones repeatedly into this cold layer of air. Tornadoes generally form from the rotating updraught at the core of a supercell thunderstorm.

Figure 2.23 Wind shear is extremely hazardous to aviation, especially during landings when pilots have difficulty making corrections to flight-path changes caused by the shifting airflow. Because of the numerous air fatalities it once caused, a wind-shear EWS was developed jointly by the FAA and NASA to warn pilots about the presence of wind shear, especially upon approach for landing *Source*: NASA.

Strong winds in thunderstorms occur not only due to tornadoes, but also due to downbursts – air that rushes rapidly downward, and then continues along the ground. Sometimes called "straight-line" winds to differentiate them from rotating tornadic winds, downbursts can reach speeds of up to 162 k/h (100 mph), and are in fact responsible for more thunderstorm wind damage than tornadoes.

Fire Danger Index (FDI)

Natural Resources Canada

The Fire Danger Index is a relative index of how easy it is to ignite vegetation, how difficult a fire may be to control and how much damage a fire may do (table 2.1).

The Canadian Forest Fire Weather Index (FWI) system is a model that predicts the potential fire danger within a forest. Such a prediction is achieved by modelling the dryness of the forest fuels (those parts of the forest that burn) and the potential of fire behaviour (how fast and intense a fire may become) (*Weather Network*, 2007).

The FWI system consists of six components that together account for the effects of fuel moisture and fire behaviour. The first three are codes that quantify moisture contents.

- *Fine Fuel Moisture Code (FFMC).* A numerical rating of the moisture content of litter and other cured fine fuels. Such fuel consists mostly of dead and fallen needles and leaves, as well as lichens, mosses and other small, loose debris. The FFMC is an indicator of the relative ease of ignition and flammability of fuel.

- *Duff Moisture Code (DMC).* A numerical rating of the average moisture content of the duff layers, a stratum of loosely compacted organic material of moderate

Table 2.1 Fire Danger Index (FDI)

LOW	Fires likely to be self-extinguishing and new ignitions unlikely. Any existing fires limited to smouldering in deep, drier layers
MODERATE	Creeping or gentle surface fires. Fires easily contained by ground crews with pumps and hand-tools
HIGH	Moderate to vigorous surface fire with intermittent crown involvement. Challenging for ground crews to handle; heavy equipment (bulldozers, tanker trucks and aircraft) often required to contain fire
VERY HIGH	High-intensity fire with partial to full crown involvement. Conditions at head of fire beyond the ability of ground crews; air attack with retardant required to control fire's head effectively
EXTREME	Fast-spreading, high-intensity crown fire (a forest fire that advances with great speed, jumping from crown to crown ahead of the ground fire). Very difficult to control. Suppression actions limited to flanks, with only indirect actions possible against the fire's head

depth on the ground. The DMC provides an indication of fuel consumption in the moderate duff layers and in large logs.

- *Drought Code (DC).* A numerical rating of the average moisture content of deep, compact organic layers on the floor of the forest. The DC is a useful indicator of seasonal drought effects on forest fuels and of the amount of smouldering in deeper duff layers and large logs.

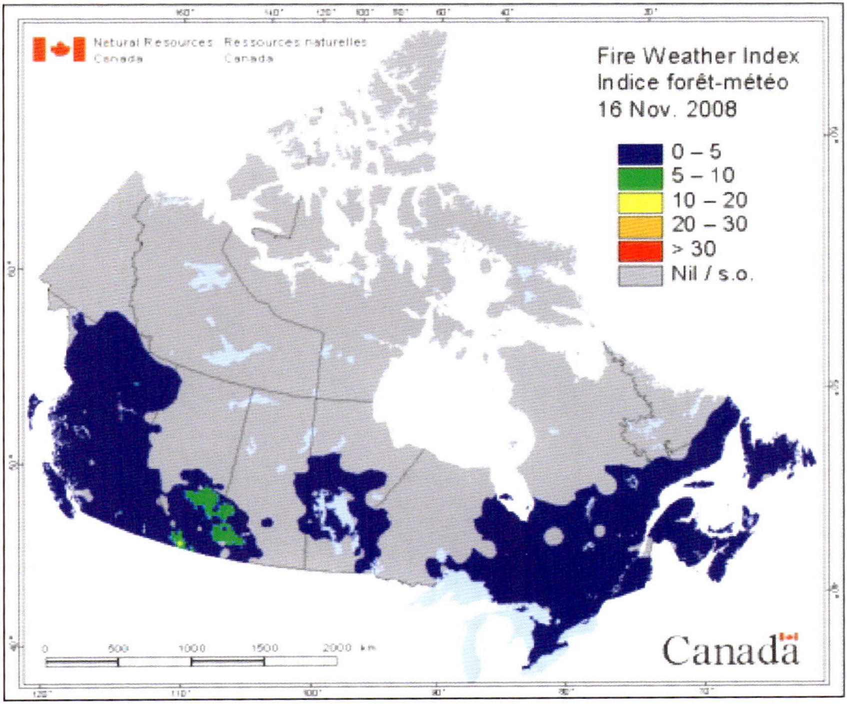

Figure 2.24 Fire Weather Index (FWI) for Canada
Source: Natural Resources Canada.

The remaining three components are fire behaviour indices that represent the rate of fire spread, the fuel available for combustion and the frontal fire intensity. Their values rise as the fire danger increases.

- *Initial Spread Index (ISI).* A numerical rating of the expected rate of fire spread. It combines the effects of wind and the FFMC without the influence of different fuel types.

- *Build up Index (BUI).* A numerical rating of the total amount of fuel available for combustion (combining the DMC and the DC).

- *Fire Weather Index (FWI).* A numerical rating of fire intensity that combines the ISI and the BUI. It is suitable as a general index of fire danger in forested areas throughout Canada.

Anatomy and Australian bushfires

Bureau of Meteorology, Australia

The big killer in a bushfire is heat radiating from the fire on to bare skin. You need to cover all bare skin with a long-sleeved shirt; long trousers, jeans or overalls; sturdy shoes; and wool or cotton socks. If you are outside, don't forget to wear a wide-brimmed hat or hard-hat, goggles or glasses, and gloves.

Other causes of death are inhalation of super-hot air and flames; dehydration in the hot, dry conditions; and being hit by falling trees and branches. Although radiant heat is fierce, it is easily blocked by a solid barrier such as the wall of a building. If you are caught in a bushfire, the safest place is inside a building away from the radiant heat, hot air and flames.

Figure 2.25 Pine forest fire in Canada
Source: Parcs Canada.

Peak values of radiant heat from even the worst bushfire rarely last more than 10 minutes. If you can shelter for this time, you have a good chance of survival. Houses have never been known to burn down in the first 10 minutes when the fire front is moving past; they are usually ignited by blowing embers, radiant heat and the burning of adjacent structures like wood heaps, trellises and timber decking, and they can take hours to burn down. If you shelter inside while the fire front passes and then put out the small spot fires inside and around the outside of your house, there is an excellent chance that you and your home will survive.

The dry and hot air around bushfires can cause dehydration. Remember to have a drink of water as often as you can, preferably every 10 minutes. Do not drink alcohol; it will dehydrate you even more.

On fire danger days, use the radio to keep up with the latest information on the weather and fire situation. If you plan to stay in your house, make sure you are well prepared. If you plan to leave, then leave early. Getting trapped by a fire when you are out in the open is very dangerous. Keep a woollen blanket in the car and get under it if you are trapped in a fire. Remember that car petrol tanks don't usually explode in a bushfire; you are much safer in your car than out in the open, and you are much safer in a building than in a car. Further information is available at www.bom.gov.au/inside/services_policy/fire_ag/bushfire/protect.htm.

Australia is plagued by the constant fear of bushfires, especially in urban areas (fig 2.27). One of its worst bushfires, named the Ash Wednesday Fires, threatened towns near Melbourne in 1983. Major bushfires also threatened the outskirts of Sydney in late December 2001 and again in January 2002. Although Australia has developed sophisticated plans to reduce fire risk, these fires were so large and widespread that they outmatched firefighters' ability to cope. Unfortunately, climate change projections for Australia envision more fires, more frequently, and with greater intensity.

Figure 2.26 Satellite image of bushfires sweeping across New South Wales, Australia
Source: NASA.

Figure 2.27 A bushfire consumes outbuildings in Tasmania, Australia
Source: Huon Valley Council, Tasmania.

Air pollution in metropolitan areas

Paulette Middleton

Every year, millions of people in the United States are exposed to unhealthy, hazy, polluted air. Many urban areas throughout the country have been designated as highly polluted, and once-pristine vistas in many national parks and wilderness areas are becoming hazier and hazier. In most areas, hazy air means very unhealthy air.

Air quality indices and forecasts are already widely used to alert the public when air quality is below certain "acceptable" standards. An air quality index tells people about current conditions, while air quality forecasts let people know what air quality is likely to be in the near future. Air quality forecasts can also be used to trigger emissions management regimes, such as no-drive days, no-burn days and even, where possible, reduced operation of polluting industrial facilities. All of these actions are aimed at quickly reducing levels of harmful pollutants in the air.

- *Air Quality Index (AQI).* Reports the daily level of air pollution on an hourly basis. The index relays the highest level of carbon monoxide, fine particulates or ozone, depending on which pollutant has the greatest hourly concentration. Values greater than 100 for carbon monoxide, fine particulates and ozone indicate that pollutants exceed state and federal standards.

- *Visibility Standard Index (VSI).* Reports the air's visual quality in the Denver metropolitan area.

The Air Quality Index (AQI), based on concentrations and health impacts of several key pollutants, has been adopted throughout the United States. It is designed to help people understand what local air quality means to public health. The AQI (table 2.2) has six colour-coded categories related to levels of health concern.

Table 2.2 Air Quality Index health categories

Air Quality Index levels of health concern	Numerical value	Meaning
Good	0–50	Air quality is considered satisfactory, and air pollution poses little or no risk
Moderate	51–100	Air quality is acceptable; however, for some pollutants there may be a moderate health concern for a very small number of people who are unusually sensitive to air pollution
Unhealthy for sensitive groups	101–150	Members of sensitive groups may experience health effects; the general public is not likely to be affected
Unhealthy	151–200	Everyone may begin to experience health effects; members of sensitive groups may experience more serious health effects
Very unhealthy	201–300	Health alert: everyone may experience more serious health effects
Hazardous	> 300	Health warnings of emergency conditions; the entire population is more likely to be affected

Source: AIRNow.

Forecasts for ozone and fine particles, two major classes of pollutants responsible for unhealthy air and (in the case of fine particles) hazy air, are currently issued by over 80 state and local air quality agencies for over 275 cities throughout the United States as part of the EPA's AIRNow programme (http://airnow.gov/). These air quality agencies are the local authorities responsible for predicting and communicating (and sometimes mandated to predict and communicate) local air quality to the public, issuing health advisories and encouraging voluntary actions to reduce pollution levels.

Many cities also continuously update websites with pollutant concentration information, so actual pollution levels and potential health concerns for individual pollutants or classes of pollutants are readily accessible to citizens. Denver, Colorado, US, is unique in that it also has a visibility index.

In Denver, the High Pollution Advisory Program is managed by the Air Pollution Control Division of the Colorado Department of Public Health and Environment. Air quality advisories are issued through the programme daily at 4 pm from 31 October to 31 March.

- **RED** advisories indicate that either current air quality is poor or conditions are expected to worsen in the coming hours. Red advisories trigger mandatory wood-burning restrictions and encourage voluntary driving reductions in the six-county Denver metropolitan area.

- **BLUE** advisories indicate that air quality is good or moderate, and no restrictions are in place while the advisory is in effect.

The Denver Visibility Standard Index (DVSI) reports the air's visual quality in the Denver metropolitan area. The visibility standard is 0.076 of atmospheric extinction per kilometre, which means that 7.6 per cent of the light in a kilometre of air is blocked. Pollution levels must exceed this standard based on a four-hour average for a violation to occur. Poor-visibility days are also in the poor range for health impacts associated with fine particulates that degrade visibility.

Studies since 1997 have consistently linked ground-level ozone with increased risk of asthma in children, wheezing, sore throats, coughs, asthma attacks, decreased lung function and long-term lung damage in children. While battles over how best to improve air quality and protect clean areas from further deterioration continue, people must be protected from the effects of pollution now. The public must be informed about the status of their air, especially if conditions pose a real threat. For many individuals – particu-

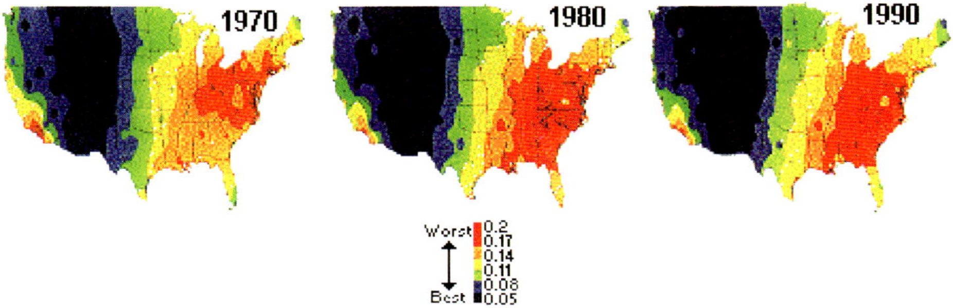

Figure 2.28 Visibility impairment in the United States
Source: US EPA AIRTrends 1996 Summary.

larly children outdoor workers and those who suffer from asthma and other respiratory problems – knowing current and forecasted levels of pollution can make a significant difference to their quality of health and well-being and in how they plan their daily activities.

Dust and sandstorm early warning

Qian Ye

Dust and sandstorms (DSSs) are natural phenomena, usually occurring in arid and semi-arid areas during dry seasons. In meteorological terms, these weather events are classified based on air visibility and wind speed as floating dust, blowing sand and sandstorms.

- *Floating Dust:* Visibility is less than 10 km but greater than 1 km, with wind speed less than 1 m/s (metres per second).
- *Blowing Sand:* Visibility is less than 10 km but greater than 1 km, with wind speed greater than 1 m/s but less than 6.5 m/s.
- *Sandstorms:* Visibility is less than 1 km, with wind speed greater than 6.5 m/s.

Dust and sandstorms occur when strong winds blow a large quantity of dust and fine sand particles off the ground and into the air. Regional weather, topography, soil properties and land use, among several other factors, affect the entrainment and transport of the particles that make up dust and sandstorms. The major sources of these phenomena are desert and semi-desert areas around the globe, such as the Sahara in Africa, the Australian desert, the Taklamakan and Gobi Deserts in China and greater Central Asia and the Loess Plateau in inland China. In these regions, dust and sandstorms are natural occurrences that have arisen for thousands of years. In China, for example, several DSS events are expected every spring, while an occasional event occurring in the fall is not considered anomalous.

Over the past 50 years, however, DSS frequencies have increased, their intensities have swelled and their geographic extents have expanded, resulting in serious social, environmental and economic impacts. Reports of very strong DSS events hitting populous urban areas have become more common, and as a result the health impacts of DSSs, which were rarely considered as part of the damage from DSS events in the past, have become one of the central concerns of the general public and governments. For example, two of the most severe Asian DSSs occurred one after another in March and April of 2002, sweeping across Mongolia and hitting 18 provinces in China as well as the Korean Peninsula and a large part of Japan. Total suspended particulate levels in the affected areas were recorded at tens to hundreds of times higher than established national standards. Such particulate concentrations have major social implications, including adverse impacts on human health as well as the increased medical costs associated with such impacts.

With the increasing rate of urbanization and the intensification of economic development, many countries in arid and semi-arid regions have also experienced various aspects of the environmental impacts of large-scale DSS events. Direct damage caused by DSSs includes loss of crops and livestock; loss of topsoil; damage to property, industries and businesses, critical facilities and infrastructure; disruption of transportation systems; road accidents; and closure of schools, production facilities and social services. The costs of cleaning residential and commercial buildings, repair and reconstruction, and devaluation caused by wear and tear on machinery and equipment have also increased over the last few years.

Figure 2.29 A sandstorm in Gansu Province, China
Source: *Gansu Daily*, May 1993.

Intensified land development in economically emergent nations like China has also increased the land area subject to DSS events because of the unsustainable land-use decisions (e.g. overgrazing, deforestation, overexploitation of water resources, etc.) that often come with robust economic growth. These decisions lead to serious environmental problems, including rapid land degradation and even desertification – the creation of desert-like conditions on once-arable land.

In recent years satellite images of dust and sandstorms and analyses of dust samples collected on the ground have revealed that the impacts of these events are not limited to their originating regions, but can reach hundreds and even thousands of kilometres away. As strong winds sweep across the growing expanses of arid and semi-arid lands in North Asia, for example, storms have sometimes become so massive and long-lasting that they have transported sand and dust across the northern Pacific, affecting not only Korea and Japan but even North America.

Because the once-regional impacts of DSSs are becoming increasingly global, with larger and larger events affecting greater expanses of the world every year, actions to manage them should be implemented that include both short- and long-term strategies. Short-term measures include forecasting and the development of early warning systems that can help decision-makers in origination countries to prepare effective protections against direct damage in the source area and in regions downwind. Long-term measures, on the other hand, require careful consideration of historical, social and ecological factors in DSS origination regions in order to prevent, for example, overgrazing and overcultivation. Mitigation activities such as replanting vegetation and land-use changes to prevent and control dust mobility must also be implemented. Furthermore, strong international cooperation is needed to implement selected DSS measures by providing sufficient financial, technical and moral support.

An illustrative case

Severe damage to humans and livestock resulted from a sandstorm in northwest China on 5 May 1993. This event directly affected 1.1 million square kilometres, and resulted in the deaths of 85 people and injury to 264 others. In addition, 370,000 hectares of agricultural land were damaged, and 120,000 domesticated animals perished. Damage to infrastructure and the manufacturing sector was reported as well. In total, the direct economic loss from this memorable event was estimated to have been about 7.3 billion yuan (about US$1 billion at 2008 rates).

3

Too much, too little

"Water is life' is a phrase that exists in one form or another in most languages. Water is and always has been, since the earliest beginnings of human societies, a central preoccupation of peoples and governments in all areas of the world. Today, we know a great deal about water and water-related issues: where it is, when it is supposed to come and from where it comes. Most of the time, however, although we know a lot about water resources in a given region, much remains that we still do not know. Some societies manage to flourish, while others struggle from year to year just to subsist on available or accessible water supplies. Ultimately, every society's greatest struggles relate in some way to finding the most effective use of its water resources.

While many believe that water is an abundant resource freely available to all societies, except those in arid lands and desert areas, the truth is that freshwater availability is relatively limited (fig. 3.1). While there are atmospheric precipitations, surface waters and groundwater resources, a major problem with water in general is its varied distribution on and around the globe. Fresh water is not readily available to all people, not only because of the aridity of some regional climates, but also because of the economic status of various segments of many populations.

Concerns about securing water supplies sufficient for agricultural, industrial and domestic purposes are exceptionally important and understandable, of course, but underlying these concerns, in every society, is the looming threat of water extremes: either too much at a given moment or too little for an extended interval – floods and droughts.

Heads Up! Early Warning Systems for Climate-, Water- and Weather-Related Hazards, Michael H. Glantz (ed.), United Nations University Press, 2009, ISBN 978-92-808-1169-8

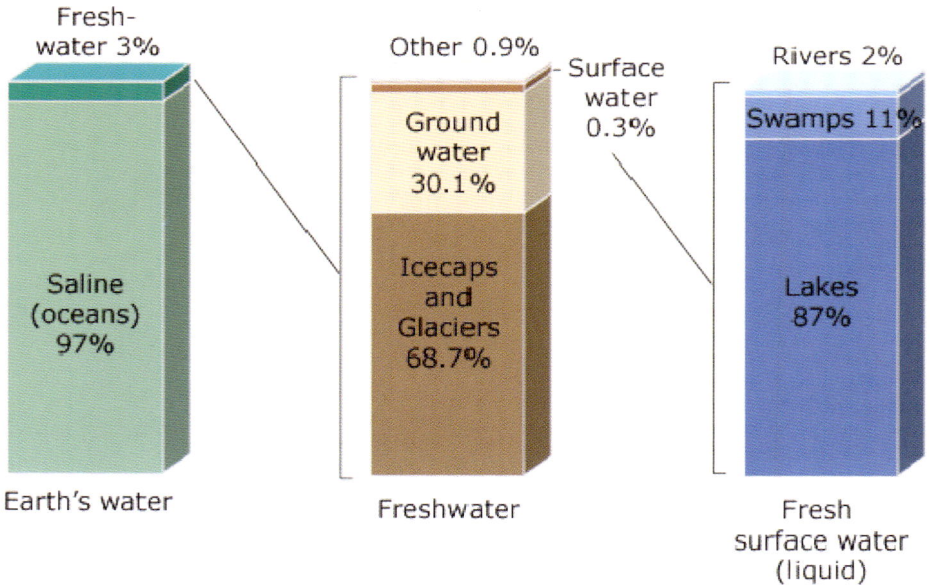

Figure 3.1 Current distribution of the earth's water. Drawn from data in Gleick (1996)
Source: US Geological Survey (2006).

Floods

In the United States, local National Weather Service forecast offices issue regular predictions for possible flooding, as well as specific forecasts for individual hydro-meteorological events that are expected to result in localized flood events. The EWS is standardized nationally.

- *Significant Flood Outlook:* Intended to provide a general outlook for significant river flooding. Most people likely to be affected by flooding are aware to various extents of flooding possibilities along rivers and creeks and in low-lying coastal and inland areas, so in a sense they are already warned of the risk of flooding once heavy rains begin at some location upstream and are reported by the news media.

- *Public Advisories:* Issued to provide information to the public about:
  - height of a flood crest
  - date and time a river is expected to overflow its banks
  - date and time a river is expected to recede back within its banks.

- *Flood Watch:* Issued when the high flow or overflow of water from a river is possible in a given time period. Watches are generally issued when flooding is expected to occur at least six hours after rain has stopped.

- *Flood Warning:* Issued when flooding conditions are actually occurring or are imminent in the warning area.

n addition, an indication of the likelihood of flood is issued.

- *Possible* – Hydro-meteorological conditions indicate that significant flooding could occur, but such flooding is neither certain nor imminent.

- *Likely* – Hydro-meteorological conditions indicate that significant flooding can be expected during the outlook period.

- *Imminent or occurring* – Significant flooding is already occurring or is forecasted to occur during the outlook period.

Floods occur somewhere around the globe each and every year; however, not all floods are alike. Some are quick onset with little lead-time for warning, while others build up incrementally in stages over a matter of days, so effective warnings about them can be issued and their adverse impacts can be anticipated. In some locations floods are rare events, while other locations can be viewed as flood-prone or highly susceptible to heavy rains and severe flooding.

Figure 3.2 As the water recedes, traffic signs begin to reappear in East Grand Forks, Minnesota, US, in April 1997
Source: David Saville/FEMA.

Whether they occur in industrialized nations or in agricultural ones, floods are devastating to lives and property. They can be regular seasonal occurrences, as in Bangladesh, or they can be quasi-periodic, occurring only every few years under the right climate conditions, like those associated with the extreme air-sea interactions of an El Niño event. Regardless, most countries have had a memorable flood event. For example, in 2000 Mozambique suffered one of its worst floods, the horror of which was captured in an image on South African TV of a woman giving birth in a tree, the swirling floodwaters lapping from below. In Caracas, Venezuela, in 1999 heavy rains, floods and ensuing mudslides were blamed for more than

Figure 3.3 A river in Montana, US, overflows its banks, washing out a road
Source: National Weather Service, Billings, Montana, Forecast Office.

50,000 deaths In the United States, the devastating Mississippi River floods of 1993 and 2008 were revelations to government agencies: channelling the river over many decades had put communities developed on the natural floodplain at greater risk than if room had been made for the river's waters. The list of floods is endless, which is why monitoring the factors that can lead to flooding and feeding that information into national early warning systems can reduce losses of life and property in flood events in most countries.

China's 1998 Great Yangtze River Floods

Qian Ye

In the summer of 1998, China experienced massive flooding along parts of the Yangtze River, the longest river in Asia. Due to heavy rainfall during June and July, the river crested in August to its highest levels since 1954. This caused several major levees along the river to collapse (fig. 3.4), resulting in more than 3,000 deaths, 14 million left homeless and US$24 billion in economic losses.

The abnormal climate in 1998, especially the occurrence of one of the strongest El Niño events of the twentieth century, was initially blamed as the main cause of the catastrophe. Follow-up studies, however, showed that inappropriate land use along the river played a significant role in contributing to flood damage. For example, deforestation in much of the Yangtze watershed caused large quantities of topsoil to erode and wash into the river. Over several decades, the build-up of this soil had raised the Yangtze's river bottom, helping to push floodwaters over the river's banks. Furthermore, wetlands and lakes in east-central China that once provided a buffer to Yangtze overflow by storing excess rainfall had been drained to make room for agricultural expansion.

After the Great Yangtze River Floods of 1998, several assessments and other studies were conducted which found that, although China was quite capable of responding effectively to natural disasters, the central government lacked science-based early warning systems to help protect the society against such disasters. Some major lessons were also identified. For example, creation of a reliable climate-, water- and weather-related early warning system, development of a high-quality scientific research programme and continued

Figure 3.4 Chinese military personnel trying to save a levee during the 1998 Yangtze River flood

monitoring to obtain forecasts with reduced uncertainties were needed. Also, better communication channels between government representatives and scientists about the state of climate science and the limits of predicting natural phenomena such as ENSO-related extreme events had to be established. Lastly, an organization was needed that could better coordinate various government agencies' responses to natural disasters.

In China there is a four-level response system for floods, emanating from the Office of State Flood Control and Drought Relief.

- *Level 1.* The most severe situation. When a serious flood is occurring in one major river basin or several large-scale floods are occurring in several river basins or in several provinces, a national consultation conference is held with the head of the State Flood Control and Drought Relief Headquarters (SFDH), all relevant central government agencies and provincial governments to develop appropriate responses. The results must be reported to the State Council and a national emergency response action may be issued. Warnings to the general public are issued on national TV every day.

- *Level 2.* When a large flood is occurring in one major river basin, dikes are broken along major rivers and/or several major cities are experiencing flooding, the deputy head of the SFDH will lead an internal meeting to discuss appropriate measures to address the situation. The results are to be reported to the State Council within two hours. Warnings to the general public are issued as necessary.

- *Level 3.* When a potentially severe flood is forecast for one major river basin or in several provinces, the head of the SFDH leads an internal meeting to review the situation and reports to the State Council within two hours. No public warnings are issued.

- *Level 4.* When floods are occurring in major river basins or in several provinces, the office director of the SFDH leads a meeting to summarize the situation and reports results to the State Council and other government agencies. No public warnings are issued.

Flash floods

Eve Gruntfest

A flash flood is defined by the US National Weather Service (NWS) as a flood resulting from heavy or excessive rainfall in a short period of time and occurring generally within six hours of the causative event. Deadly flash-flood-hazard areas are often found in arid and semi-arid climates, steep topography or places where infrequent precipitation comes in the form of intense thunderstorms.

In terms of *flash-flood awareness*, general knowledge and information are available in the United States about locations prone to flash-flooding preconditions. For example, people are warned not to picnic in dry river-beds, and those who live in narrow mountain basins are warned to move to higher ground in the event of heavy rains. Signs are often posted in areas prone to flash flooding.

Public advisories take two forms.

- *Flash-Flood Watch:* Flash flooding is possible in or close to the watch area. Generally issued when flooding is expected within six hours after a storm ends.

- *Flash-Flood Warning:* Flash flooding is occurring or is imminent in the warning area. Issued as a result of torrential rains capable of causing flooding as observed by radar and/or satellite, dam or levee failure, or ice jam.

Figure 3.5 Fort Collins, Colorado, US, flood of 1997
Source: John Weaver/NOAA.

Urban and *small-stream flood advisories* are issued when minor widespread urban flooding along small streams is occurring or imminent. Though streets, low-lying areas, storm drains and underpasses are flooded, this type of flooding is generally only an inconvenience and not life-threatening.

Official warnings do not always precede flash floods, especially in small watersheds. Often, environmental cues, including increased noise from a stream, extremely heavy rainfall or roads blocked by rocks or trees, are the only warnings. On the Internet, flash-flood threat maps are updated in real time, showing the amount of rainfall within a three-hour period across the United States that could cause flooding (Gruntfest and Ripps, 2000).

In developing and developed countries alike, flash floods are often quite devastating, especially when they trigger mudslides. In early January 2007, for example, flash floods in Rio de Janeiro, Brazil, killed 31 people when mudslides after heavy rains from a severe weather system rapidly filled rivers. Another deadly example was when 17 people in southern California were killed on Christmas Day 2003 after flash floods caused severe mudslides. The disaster occurred on hillsides where buildings and vegetation had been destroyed by massive wildfires only two months earlier.

In the United States, where most fatal events take only a few lives at a time, an average of 100 deaths each year result from flash floods, the majority of which are caused by individuals driving into flooded roadways. One key problem is that people consistently underestimate how much water it takes to lift a car from the road. Many do not realize that a vehicle can be lifted off the ground by as little as 1.5 feet (0.5 metres) of water. Only when people are convinced that "it is better to be wet than dead" and to take flash-flood warnings and environmental cues seriously will the number of flash-flood fatalities decrease in America (ibid.). The last flash flood to kill more than 100 people in the United States was the Big Thompson flood in Colorado in 1976.

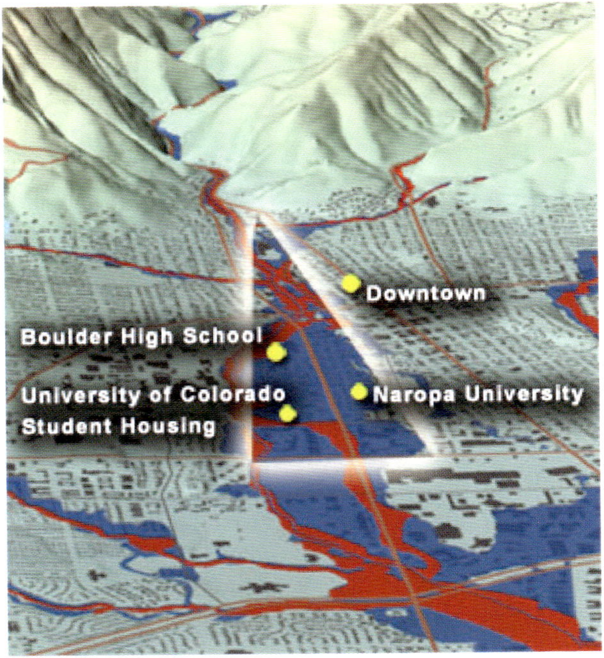

Figure 3.6 "It's a matter of when" in especially vulnerable Boulder, Colorado, US. The area in blue shows the "100-year floodplain". The area expected to have the deepest and most powerful current, cutting through densely populated and trafficked areas, is in red

Source: www.boulderfloods.org.

Urbanized areas are increasingly affected by the deadly threat of flash floods because unwise land-use decisions and urbanization have increased exposure and vulnerability (see fig. 3.6 for an example). Urbanization has especially led to increased run-off speeds, which amplify flash-flood risk. Overgrown vegetation, inadequate drainage and under-designed culverts may also lead to flash flooding.

Unfortunately, exposure and vulnerability are increasing as populations swell and more and more people move into flood-prone areas. The best detection system alone will not save lives. Messages must be delivered in a timely fashion, and people at risk must know what actions to take and take them promptly. Even with state-of-the-art technology for radar and gauge data and professional staff hard at work, small, highly localized flash floods will always prove difficult to detect and to warn for, potentially resulting in a loss of life.

Droughts

Droughts are occurring in various locations on the globe all year round. There are three different types of drought: meteorological, agricultural and hydrological. Each has its own kinds of impacts on society and the environment.

Meteorological drought is an expression of a deficiency of precipitation from normal or expected amounts over some extended period of time. Agricultural drought, on the other hand, is more closely linked to deficiencies of moisture in the soil root zone, which is the primary determinant of plant growth. In contrast, hydrological drought is associated with insufficient water supplies, both in surface (e.g. reservoirs, stream flow and snow pack) and subsurface (e.g. groundwater, aquifers) systems. All three types of drought vary in time and space, with some episodes lasting for several weeks within a single season and others lasting for several consecutive years. Because droughts typically result from multiple factors (not just from a lack of rain), determining the precise beginning and end of a drought episode is often quite difficult.

Monitoring for drought, therefore, is not an easy task because dry spells and droughts, like some other natural hazards, are slow-onset, cumulative, creeping phenomena. Consider the following account from American meteorologist I. R. Tannehill (1947):

> We welcome the first clear day after a rainy spell. Rainless days continue for a time and we are pleased to have a long spell of such fine weather. It keeps on and we are a little worried. A few days more and we are really in trouble. The first rainless day in a spell of fine weather contributes as much to the drought as the last, but no one knows how serious it will be until the last dry day is gone and the rains have come again.

Monitoring for drought is also difficult because of the large number of potential indicators necessary to describe an episode's severity (e.g. precipitation, snow pack, reservoir and lake levels, stream flow, soil moisture, groundwater levels, etc.).

Many economic sectors, such as agriculture, water supply, energy production, transportation, manufacturing, recreation and tourism, are also directly or indirectly affected by drought. A number of countries even consistently monitor for drought conditions in other parts of the world where economic competitors (regarding, for one example, agricultural production) might be experiencing drought conditions which will be likely to affect global food prices.

More worryingly, drought also results in serious social and environmental impacts that often affect the quantity and quality of food supplies, and therefore the food security of a country or an entire region. As a result, untrained observers typically conclude that drought causes famine; however, closer scrutiny of a famine will in most cases reveal a multi-stressed environment in which the drought itself was only one of the pressures that led to the famine – and not necessarily the most important one.

Figure 3.7 The drought in Spain in 2005 was the worst the country has ever experienced
Source: WMO.

Figure 3.8 Women spend hours gathering water in Ethiopia
Source: M. H. Glantz.

In an industrialized country, food insecurity can certainly cause significant economic problems; in a developing country, however, a drought, especially an agricultural drought, can become a matter of life, sickness and death. Because of the complex factors intrinsic to drought episodes, developing an effective and comprehensive national early warning system for drought is difficult, although nations have created such systems in the hopes of giving a collective "heads up" on impending food shortages, price increases for agricultural products, food export and import problems and impacts on trade.

NOAA's Palmer Drought Severity Index and Crop Moisture Index

The Palmer Drought Severity Index (PDSI) was developed in the 1960s by meteorologist Wayne Palmer, who used temperature and rainfall information in a formula to determine dryness. It has become the semi-official US drought index.

In the index, "0" indicates normal conditions, while drought conditions are reflected in terms of negative numbers. For example, negative 2 indicates moderate drought, negative 3 indicates severe drought and negative 4 indicates extreme drought. The Palmer Index can also reflect excess rain using positive integers (i.e. "0" is normal, positive 2 indicates moderate rainfall, etc.).

The PDSI is most effective in determining long-term drought (a matter of several months); its accuracy diminishes with short-term forecasts (a matter of weeks). The advantage of the PDSI is that it is standardized to local climates, so it can be applied to any part of the country to demonstrate relative drought or rainfall conditions. The main disadvantage, as mentioned, is its inaccuracy for short-term forecasts, so it works best east of the Continental Divide.

The Crop Moisture Index (CMI) is a formula developed by Palmer subsequent to his development of the PDSI. Using the same scale as the PDSI, the CMI is designed to

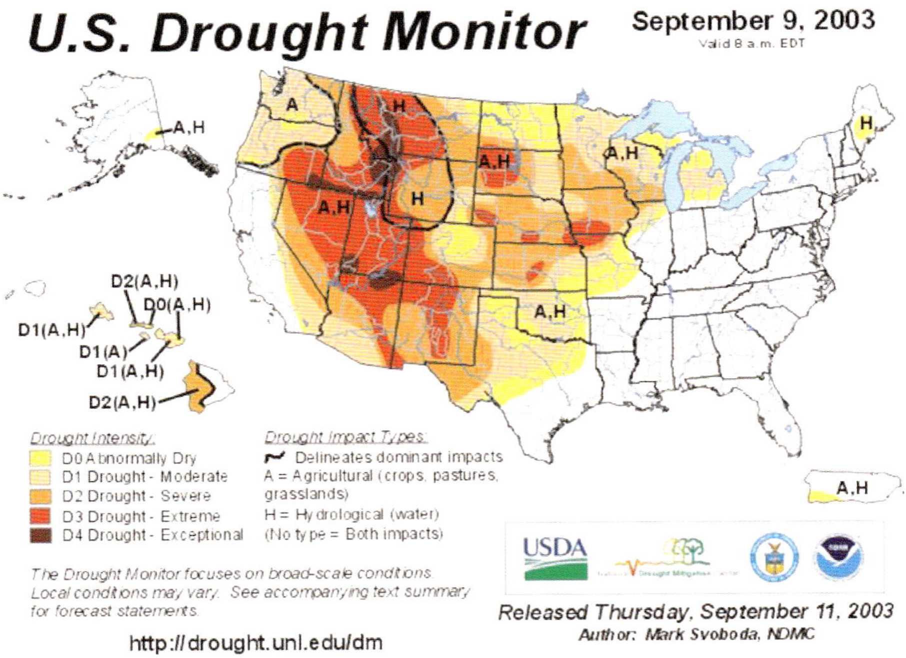

Figure 3.9 The US Drought Monitor is developed weekly through a partnership between the National Drought Mitigation Center, the University of Nebraska-Lincoln, the National Oceanic and Atmospheric Administration and the US Department of Agriculture. It has become the primary early warning tool for drought in the United States, identifying both the extent and the severity of drought conditions on a weekly basis

indicate normal conditions at the beginning and end of the growing season, differing from the PDSI in that the formula places less value on data from previous weeks and more on those from the most recent week. It responds more rapidly than the Palmer Index and can change considerably from week to week, so it is more effective in determining short-term abnormal episodes of dryness or wetness that might affect agricultural production.

Drought in Africa

Stefanie Herrmann

With over 40 per cent of the African continent categorized as drylands (not counting hyper-arid areas) on which rainfall variability is inherently large at annual to decadal time scales, drought is a common occurrence. In fact, droughts of varying lengths and intensities have been observed since the beginning of the historical period for which rainfall records are available. In response, most indigenous agricultural and pastoral land-use systems in the African drylands have developed around strategies, including mobility, diversification and risk spreading, to cope with rainfall variability and drought. Nevertheless, drought events, particularly severe ones, do cause considerable hardship for affected populations (see, for example, fig. 3.10). Droughts have had dramatic impacts on both the environment and the economies of drought-stricken regions, and have led to the periodic disappearance of villages and even whole ways of life. With the eventual return of good rainfall years, however, the environment recovers and humans return to resettle abandoned regions, rebuild herds and again prepare the land for cultivation.

Beginning in the late 1960s, an unusually long period of below-average rainfall hit the Sahel zone, a semi-arid region stretching along the southern edge of the Sahara Desert. Negative rainfall anomalies prevailed until the early 1990s and were accompanied by devastating land degradation (e.g. desertification), food insecurity and famine. The few slightly better rainfall years that occurred during this period (e.g. 1975, 1978 and 1988)

Figure 3.10 Carcasses of dead cows after a drought in Kenya
Source: Médecins sans Frontières.

were too brief for either the environment or the population to recover fully. Moreover, continuously growing populations as well as government restrictions on pastoral mobility increased pressures on resources and undermined traditional drought adaptation mechanisms. While the case of the Sahel presents the most dramatic example of drought and its impacts in Africa, droughts have repeatedly made headlines throughout northwest, East and Southern Africa as well as in the Horn of Africa in recent years.

The role of drought early warning in Africa

Droughts are creeping natural hazards. They often start unnoticed and develop slowly. By the time a drought (as opposed to a dry spell) becomes evident, it is often too late for farmers, pastoralists and humanitarian agencies to take mitigating action.

With the exception of South Africa and some fast-growing economies in North Africa, Africa is truly part of the developing world. A majority of its population lives below the poverty line and has insufficient access to infrastructure, and is, therefore, particularly vulnerable to climate-, water- and weather-related hazards. African livelihoods are predominantly agricultural and are thus directly dependent upon the vagaries of climate, especially since most African agriculture for both subsistence and market production is rain-fed, even in areas that are, because of low average annual rainfall, considered marginal for this type of land use. As a result, the economic and social effects of droughts are often ruinous, contributing to food insecurity and frequently to famine.

Early warning systems are crucial for enhancing drought preparedness and facilitating drought mitigation. Timely information about the onset of a drought can help farmers make crop choices, pastoralists make livestock marketing decisions and humanitarian organizations arrange to deliver direct and timely assistance to the most vulnerable populations before lives are lost.

Most drought-prone countries in Africa have regional and national food security early warning systems. In addition, NGO- and donor-operated EWSs, such as the FEWS Network, also operate at subnational to regional scales. The tools these systems use include satellite remote sensing, ranging from simple indicator measurements to complex indices that monitor climate and vegetation conditions (fig. 3.11), and ground-based social assessments such as market trends and household vulnerability assessments.

The number of formal and informal early warning systems in Africa is evidence of the growing recognition of the value of early warning; however, only a few of these systems are fully operational, and of these even fewer have been exemplary in the drought responses they have prompted. This lag can be attributed in part to the uncertainties that surround climate processes and the shortcomings of drought monitoring capabilities. Although significant advances that could one day help in prediction have been made in determining possible causes of African rainfall variability and drought, some large-scale connections are still not well understood.

Another reason for this lack of effectiveness can be explained by deficiencies in the dissemination of early warning information and in the design and implementation of effectual response strategies – the ultimate goal of a holistic early warning system. As noted in the Introduction of this book, depending on the set of drought indicators used, different warning systems sometimes generate contradictory information. This result can confuse decision-makers, among others, and delay timely response. Moreover, the language of probabilities used in early warning does not translate well to the type of information that an affected population can easily understand and readily use.

Famine Early Warning System Network

The Famine Early Warning System Network (FEWS-Net) is an early warning system implemented by the US Agency for International Development (USAID) in cooperation with

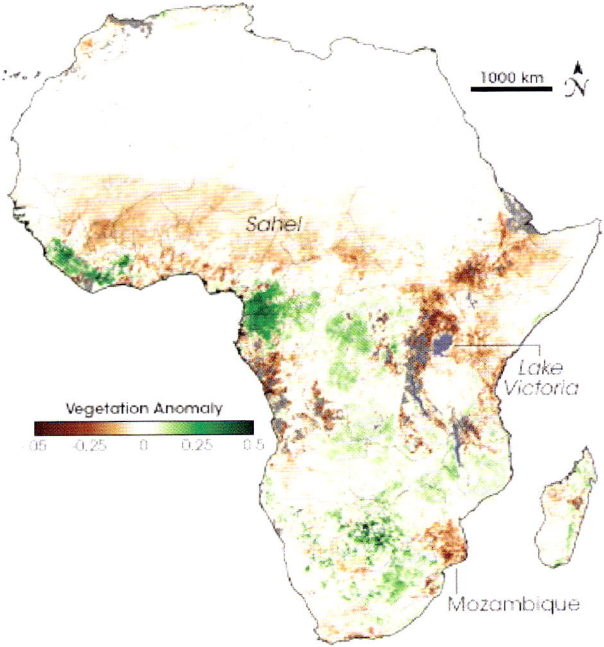

Figure 3.11 Anomalies in the Normalized Difference Vegetation Index (September 2002) show drought conditions from West Africa to the Horn of Africa. Brown indicates negative anomalies, with vegetation less than average over a 20-year period

Source: NASA.

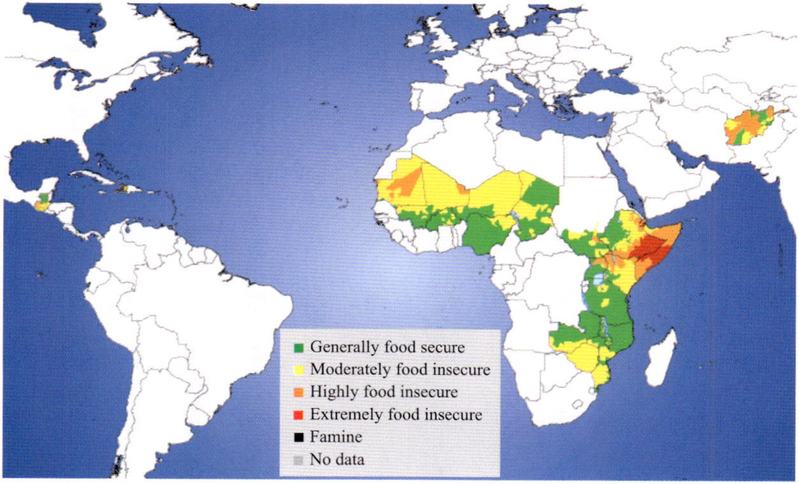

Figure 3.12 Food security conditions in several countries in the second quarter (April–June) of 2008
Source: FEWS-Net.

other American and African institutions. FEWS-Net began operating in the mid-1980s in response to large-scale droughts and famine in the Sahel. It currently provides drought and food security information for 17 African countries (fig. 3.12).

FEWS-Net issues *alerts* to decision-makers when a food security crisis begins to emerge. As the crisis deepens, these alerts are upgraded from *watches* to *warnings* to *emergen-*

cies, indicating an increasingly urgent need for action. FEWS-Net alerts also facilitate comparison of the severity of crises in different countries, thus helping to prioritize assistance. In addition to alerts, FEWS-Net releases *monthly reports* for each region and country centre as well as *special reports* as necessary.

To assess current situations and for early detection of drought hazards, FEWS-Net makes extensive use of satellite-based information (rainfall estimates, vegetation and water requirements, satisfaction indices, etc.), ground monitoring and livelihood analyses to judge better the impacts of climatic shocks on household food access. GIS is used to integrate different indicators efficiently.

FEWS-Net alert levels

- *No Alert:* No indications of food security problems.
- *Watch:* Indications of an emerging food security crisis. Decision-makers should pay increasing attention to the situations highlighted in this watch and update preparedness and contingency planning measures.
- *Warning:* A food crisis is developing in which groups are now or will soon become highly food insecure and will take increasingly irreversible actions that undermine their future food security. Decision-makers should urgently address the situations highlighted by the warning.
- *Emergency:* A significant food security crisis is occurring where portions of the population are now or will soon become extremely food insecure and face imminent famine. Decision-makers should give the highest priority to responding to situations highlighted by this alert (fig. 3.13).

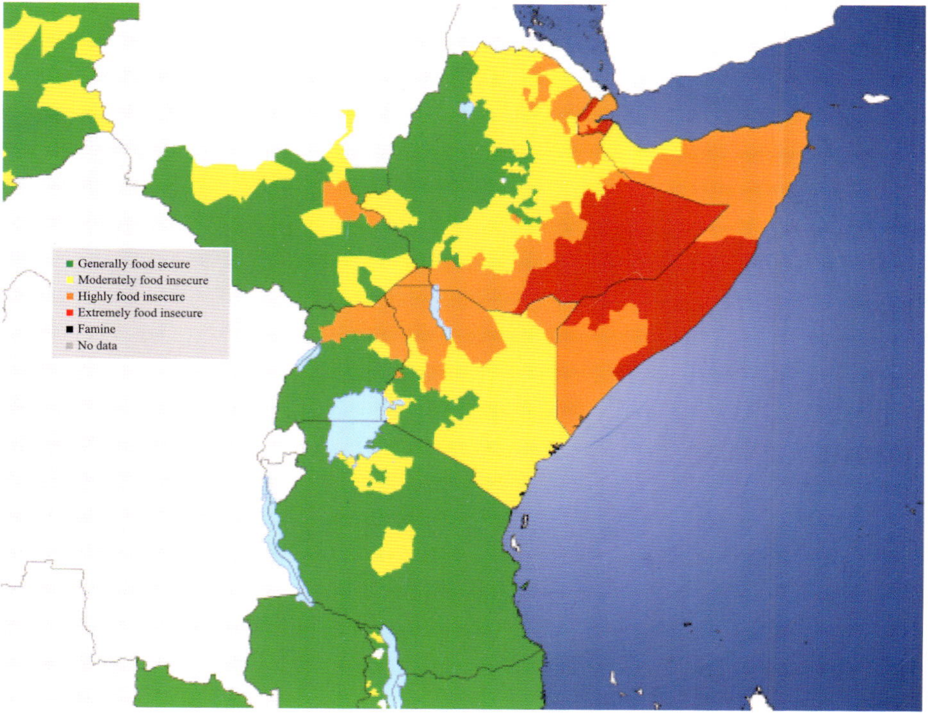

Figure 3.13 Several countries in East Africa are in emergency alert status as of June 2008
Source: FEWS-Net.

Australian monitoring for drought early warning

Bureau of Meteorology, Australia

The island continent of Australia straddles the southern subtropical zone, its mainland extending from around 11°S across the "top end" to 39°S in the southeast. The northern regions are seasonally tropical, while the eastern, southeastern and southwestern coasts and the near inland regions are generally well-watered but prone to high inter-annual and seasonal rainfall variability. Regions further inland range from semi-arid to arid. Droughts, sometimes covering vast tracts of the continent, are a recurring feature of Australia's climate; many of the more severe and widespread droughts are associated with El Niño events.

Given that rainfall is by far the dominant factor in determining the success or failure of the growing season across Australia, the monitoring of drought has for many years been synonymous with the monitoring of rainfall deficiencies. The Australian Bureau of Meteorology's Drought Watch System, in operation since 1965, uses percentiles of accumulated rainfall over successive months to identify regions of both rainfall deficit and rainfall excess. Areas with rainfall accumulations below the tenth or fifth percentile for a period of three months or more are referred to as being *seriously* or *severely* in deficit, respectively.

While an extended period of rainfall deficiency in any area is virtually a prerequisite for drought, there is widespread recognition in Australia that the formal declaration of a "drought" is a more complex issue, involving not only consideration of the supply of rain-water but also of the subsequent uses of that water once it has fallen on to farmlands, run into streams and rivers, been stored in dams and been used to drive hydroelectric stations to supply power to cities and towns. Furthermore, given the size and geographic location of Australia, it is unusual for there *not* to be one or more areas of varying size at any given time experiencing serious or severe rainfall deficiencies. Deciding whether or not such areas are declared drought stricken and then whether the drought is of sufficient intensity in duration and extent for those affected to be eligible for government relief involves a complex series of assessments by national and state authorities. The recognition that drought is a "normal" feature of Australia's natural, economic and social environments has led national and state governments to agree that climate-sensitive industries and enterprises must learn to manage drought risk along with all the other attendant and ongoing risks they face. Nonetheless, national, regional and local governments do recognize that, from time to time, some droughts become so severe, chronic or widespread that there is a need to offer support to those worst affected. Such occurrences in Australia are called "exceptional circumstances".

In 2002–2003, Australia experienced an especially severe and widespread drought (fig. 3.14) that was accompanied by record high temperatures in many regions. At the peak of the drought, 57 per cent of the Australian mainland was registering 10 months or more of serious to severe accumulated rainfall deficits, with 90 per cent recording below the median.

With the experience of this drought fresh in mind and recognizing the need for a more objective, fair and transparent process underpinning the declaration of "exceptional circumstances", the Primary Industries Ministerial Council of Australia in 2005 commissioned the establishment of the National Agricultural Monitoring System (NAMS).

Collectively, the information in NAMS indicates current conditions for major agricultural production systems and prospects for production during the upcoming growing season. While NAMS has initially been directed at monitoring and supplying data for dryland/broad-acre industries, the system is to be extended to cover the extensive irrigated regions of Australia as well as the more water-intensive industries like horticulture.

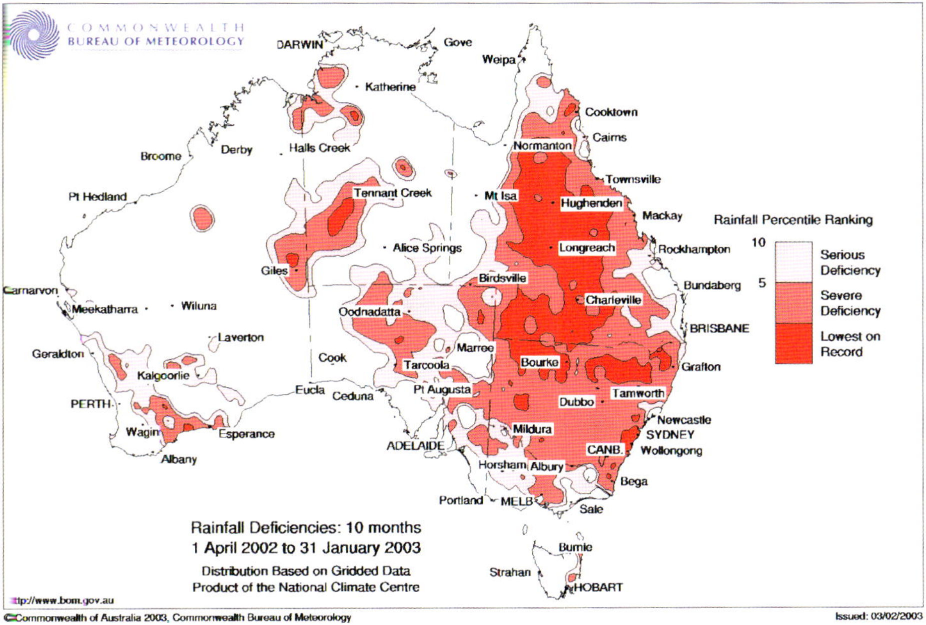

Figure 3.14 Serious or severe rainfall deficiencies at the peak of the 2002–2003 El Niño-related drought are highlighted in shades of red
Source: Australian Bureau of Meteorology.

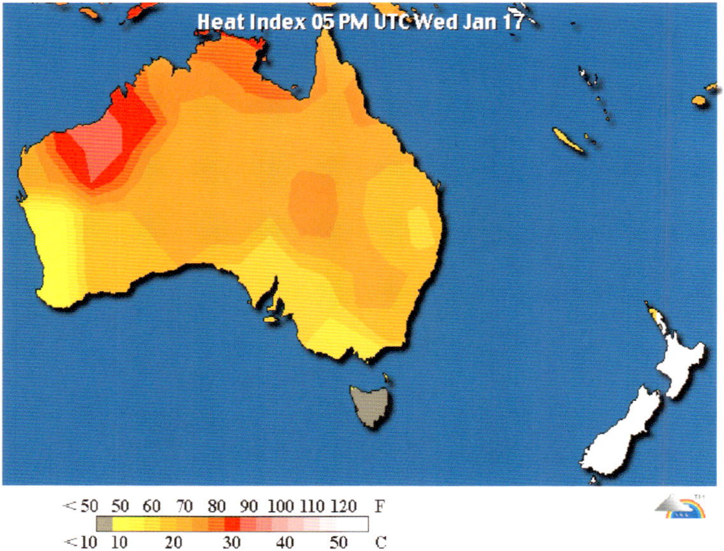

Figure 3.15 A heat index is provided on a daily basis as a public service in Australia
Source: http://wvec.com.

Because NAMS draws on a common information database for the whole country, it facilitates a more consistent approach to the drought declaration process by using:

- a common template and language for describing drought in terms of probabilities
- a common set of declaration criteria

- a common process for subjective, “on the ground” assessment of impacts.

(National Agricultural Monitoring System, 2005)

Drought in Latin America

UN Food and Agriculture Organization

Latin America is affected by natural hazards, several of which—droughts, floods, hurricanes, disease outbreaks, mudslides, El Niño and La Niña events and global warming—are related to climate, water and weather. Volcanoes are also active in Latin America and can influence climate depending on the magnitude of their eruptions. In general:

> Where El Niño produces droughts, the impact is only negative: reduction in water availability for irrigation, delayed sowing and lower yields, reduced productivity of natural grasslands and rainfed crops, increased wind erosion in flat and highland areas… the lowering of water tables and the drying up of wells, and the advance of desertification in arid, semi-arid, and sub-human ecosystems. El Niño forecasts and seasonal outlook projections for Latin America are taken very seriously by governments.
>
> Drought is the longest-lasting form of natural disaster, occurring gradually and silently. Many countries of the region have been affected by droughts that have persisted for five or more years. Between 1979 and 1984, Northeast Brazil was hit by a devastating drought that affected more than 10 million people and caused many more to migrate. The drought of 1977 in Mexico affected 1 million hectares of cropland, half of which were almost totally destroyed, and caused the death of some 45,000 head of cattle. The drought of 1982–83 in Bolivia affected 1.6 million peasant farmers.

Figure 3.16 A prolonged drought is severely threatening the food security, health and nutritional status of children and adults in Bolivia's southeastern El Chaco region. The acute nature of drought in the region puts it at risk of a major humanitarian crisis. More information is available from the UNDP Bolivia Country Office (www.undp.org/disred/english/)
Source: UNDP.

There are a number of information systems monitoring food production around the globe. Latin America is no exception. The FAO has the GIEWS (Global Information and Early Warning System), which "issues regular reports on the crop situation, the impact of natural phenomena on the output of major commodities in different regions and countries, the food outlook, trade information, and alerts and information for El Niño, hurricanes, and other specific climate threats" (FAO, 2000).

4

Global warming

Those responsible for designing and operating early warning systems related to climate, water and weather are on a learning curve. They must remain vigilant, as climate not only varies on different time scales but changes as well. As a result, those involved in early warning activities constantly have to rethink the traditional parameters used for monitoring their hazards of concern, keeping an eye open for new factors that might strengthen or hinder their ability to provide the earliest warnings possible.

Local and regional conditions under what are considered "normal" global climate regimes – normal being based on a society's short time perspectives – are constantly varying. There have been decades and sometimes even centuries of warm global temperatures (like the Medieval Optimum and, more recently, the first half of the twentieth century) and of cold periods (like the Little Ice Age and the years from 1940 to 1970).

Today, scientists continue to generate improved scenarios that support the view that the global climate regime is warming at an accelerated pace. Compelling evidence continues to mount, reinforced by the fact that the 10 hottest years on record have occurred since 1994. Indeed, observations of climate parameters and ecological changes worldwide support the scientific consensus that the earth's climate will continue to warm through the rest of the twenty-first century and beyond. Early warning systems will be vital for humans to overcome the many climate-, water- and weather-related hazards this warming will generate.

Heads Up! Early Warning Systems for Climate-, Water- and Weather-Related Hazards, Michael H. Glantz (ed.), United Nations University Press, 2009, ISBN 978-92-808-1169-8

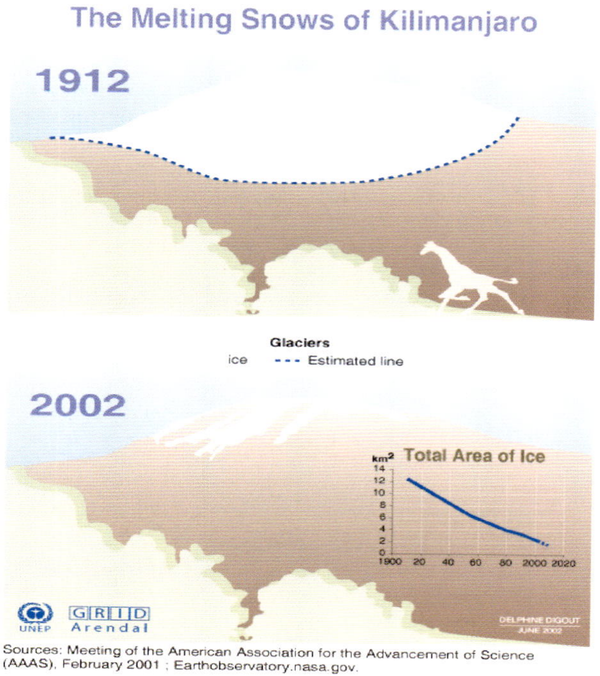

Figure 4.1 The melting snows of Kilimanjaro
Source: UNEP/GRID-Arendal Maps and Graphics Library.

Global warming

Kevin Trenberth

Planet Earth is habitable because of its location relative to the sun and because of the natural greenhouse effect of its atmosphere. Various atmospheric gases contribute to this greenhouse effect, the impact of which on clear skies is about 60 per cent from water vapour, 25 per cent from carbon dioxide and 8 per cent from ozone, with trace gases, including methane and nitrous oxide, comprising the rest. Clouds also produce a greenhouse effect.

Over the past 50 years, human influences have had a detectable influence on the earth's climate. The main way humans alter global climate is by interfering with natural flows of energy by altering the chemical composition of the atmosphere. On a global scale, even a change in energy flows of as little as 1 per cent, which is the order of the estimated change to date, will dominate all other direct influences that humans have on climate.

Global changes in atmospheric composition occur from human activities that emit greenhouse gases (GHGs) into the atmosphere. These gases include carbon dioxide that results from the burning of fossil fuels, methane and nitrous oxide. Tropical deforestation also contributes to increases in carbon dioxide levels in the atmosphere. Because GHGs have long lifespans in the atmosphere (on the order of decades to centuries), the result is an atmospheric accumulation that is clearly demonstrated by both instrumental observations and measurements of air bubbles trapped in ice cores. Carbon dioxide levels have increased 31 per cent since pre-industrial times (the early 1700s), from 280 ppmv (parts per million by volume) to over 380 ppmv today, with half of the increase having occurred since 1965 (see fig. 4.2). Greenhouse gases, acting much like a blanket, intercept and trap outgoing longwave radiation from the earth's surface on its way into space, thereby warming the planet.

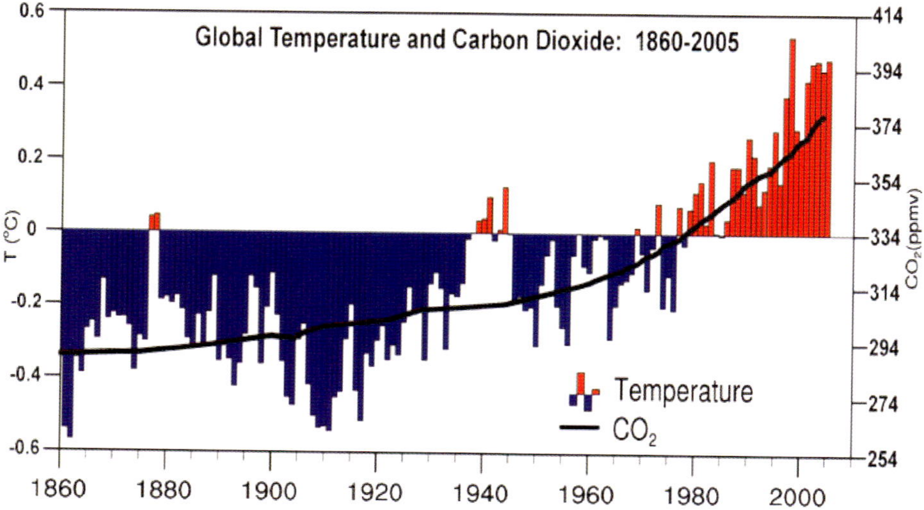

Figure 4.2 Time series of departures from the 1961–1990 base period for annual mean global temperatures (bars) of 14.0°C and carbon dioxide (solid curve with a mean of 334 ppmv during the base period) from ice cores and Mauna Loa (after 1958). The global average surface heating approximates that by carbon dioxide increases owing to cancellation of aerosols and other greenhouse gas effects, but this does not apply regionally. Many other factors, such as the effects of volcanic eruptions and solar irradiance changes, are also important

Ironically, a local cooling effect is sometimes caused by emissions from the burning of fuels that result in visible pollution (aerosols) in the atmosphere, which blocks out the sun and interferes with how clouds work; however, aerosol pollution is regularly washed or blown out of the atmosphere by precipitation events and winds, only lasting about a week or so on average, thus making it a regional rather than a global effect.

There is no longer any doubt that the composition of the atmosphere is changing because of human activities, and today greenhouse gases represent the largest human influence on global climate. Worryingly, greenhouse gas emissions in the United States and globally are increasing, up by 0.5–1 per cent per year over the past few decades. Aerosol concentrations are also estimated to be increasing, but because the effects of greenhouse gases dominate the net cooling of aerosols, the popular term for the general phenomenon of human influence on global climate is "global warming" – which is a bit of a misnomer as the effect really results in "global heating", from which observed global temperature increases are only one consequence. Rather ominously, since about 1980 the earth's climate has exceeded the bounds of natural variability.

A further consequence of the global heating of the lower troposphere is accelerated land surface drying and an increase in atmospheric water vapour, the dominant greenhouse gas. Moisture on land, if available (as it always is over the oceans), effectively acts as the "air conditioner" of the planet's surface, as heat expended in evaporation moistens the air rather than warms it. Accelerated drying increases the incidence and severity of droughts, while additional atmospheric water vapour increases the risk of heavy precipitation events. The likely results of this scenario are more frequent heat waves, droughts, extreme precipitation events and related regionally varying impacts (e.g. wildfires, heat stress, vegetation changes and sea-level rise).

Contrary to popular belief, the climate is not "changing" from one climate regime to another; rather, expectations are for a scenario of continual and unpredictable climate change into the future. Moreover, the rate of human-induced climate change is projected to occur much faster than most natural processes and certainly faster than those prevailing over the past 10,000 years. In other words, significant further climate change is guaranteed, the full impacts of which can only be estimated. The rate of change can be slowed, but is unlikely to be stopped in the twenty-first century because concentrations of long-lived greenhouse gases are dominated by the accumulation of past emissions, so many decades must pass for any change in emissions to have much effect. This means that the atmosphere still has unrealized warming (estimated to be *at least* another 0.5°C) and that sea-level rise may continue for centuries even after an abatement of anthropogenic greenhouse gas emissions and a stabilization of atmospheric gas concentrations.

Direct human influence on the global climate is not, however, limited to atmospheric emissions. Visible, large-scale human influences on land surfaces around the world through urbanization, land clearing and agricultural practices, often most pronounced where people live, work and grow food, have also had a considerable impact on climate systems. Large-scale deforestation and desertification in the Amazon Basin (South America) and the Sahel (sub-Saharan Africa), respectively, provide evidence suggesting that human alterations to land surfaces probably influence regional climates as well. Further evidence of such localized influences comes from the fact that climates in extensively developed areas, especially large urban areas with little vegetation and scores of hectares of concrete, differ considerably from those in surrounding rural, "green" areas because of the effects these "concrete jungles" have on heat retention, run-off and pollution, all of which combine to produce what are called "urban heat islands".

We are now entering the unknown with our climate. We must not only continue to monitor the vital signs of the planet, but must also assess why they are fluctuating and changing and what we must do about it. Possibly the first truly global issue in history, climate change may prove to be humanity's greatest challenge, but it is rather unlikely to be adequately addressed without greatly improved international cooperation.

The Polar Regions: A "place" as an EWS for global warming

Michael H. Glantz

"We are the first generation to influence the climate and the last generation to escape the consequences".
(US Senator John McCain, quoted in Pegg, 2004)

In discussions about global warming of the earth's atmosphere, one anecdote used quite frequently is that of the proverbial "canary in the coal mine" in reference to the impacts of global warming on the Arctic region. As the story goes, miners digging deep below the earth's surface would keep an eye on a canary in a cage they had carried down with them because canaries are especially sensitive to leaks and build-ups of noxious gases like carbon monoxide. If the canary showed signs of faltering or fell dead in its cage, it would provide an early warning to the miners to withdraw from the mine before they, too, were overcome by deadly gas or trapped by a gas explosion.

The circumpolar Arctic ecosystems (including the US state of Alaska, Greenland, Canada and the Russian Federation) and the people dependent on them for their livelihoods and well-being are like the proverbial "canary in the coal mine" for the rest of the planet because for each degree of warming in the mid-latitudes, a 3–4°C increase in temperature in the Arctic regions is expected – a difference directly resulting from the physical properties of the earth and its atmosphere. In a macro-sense, then, the Polar Regions serve the rest of the planet in the same capacity as did the canaries in the mines. In essence, therefore, all eyes should be on the circumpolar regions, as the early signs of the impacts of climate change should appear there first.

Regrettably, people who pay serious attention to human activities and ecosystems in these cold regions have been observing the early signs of global warming for some time. NASA satellite images have shown large chunks of Antarctic ice, some the size of US states, breaking away. Additionally, the area covered by Arctic sea ice has diminished considerably in the past few decades, with numerous lakes appearing where once there were only expansive ice fields (fig. 4.3). Migration patterns of Arctic animals have also shifted as the region has warmed. Most recently, global attention has focused on environmental changes that are endangering Polar bears, a fact even noted by avowed global-warming sceptic former US President George W. Bush. Indeed, just about every season a new indicator of global warming emerges from the Polar Regions.

One telling example occurred in the summer of 2005 when Canadian Ice Service analyst Laurie Weir reported that the 41 square kilometre Ayles Ice Shelf had broken away from the coast of Ellesmere Island. This event startled scientists, who had believed that ice shelves would just slowly melt away under a global warming scenario. A representative of the Inuit people in the region made the following comment: "[Inuit] find [themselves] at the very cusp of a defining event in the history of the planet. The Earth is literally melting... My Arctic homeland is now the health barometer for the planet" (Pegg, 2004).

Figure 4.3 The melting of the Polar ice

Needless to say, many projections about global warming exist; what is needed, even so, is the development of an early warning system with an inbuilt capacity to respond effectively to those projections as they become global realities. Otherwise, we will be left with yet another example of what the European Environment Agency called "late lessons from early warnings" (EEA, 2002).

Coral reef bleaching

Joanie Kleypas

The US National Oceanic and Atmospheric Administration (NOAA) uses satellite-derived sea surface temperature data to track anomalously warm water masses and identify coral reef areas threatened by these waters. They compute two values, termed *HotSpots* and *Degree Heating Week*, which can help predict coral bleaching events.

- *HotSpots:* Areas of anomalously warm water, defined simply as a region where the temperature is greater than 1°C above the maximum expected summer temperature. HotSpots indicate only the severity of a warm water event.

- *Degree Heating Week (DHW):* An index that combines the severity of a high-temperature event with its duration. For example, a DHW reading of 1.0 is equivalent to one week with temperatures elevated by 1°C above the maximum summertime temperature for a particular reef. A DHW reading of 2.0 is equivalent to one week with temperatures elevated by 2°C above the maximum, to two weeks with temperatures elevated by 1°C or to an equivalent combination of temperature and duration. And so forth.

NOAA is also installing *in situ* observation stations, called the *Coral Reef Early Warning System* (CREWS), at major coral reef regions in US waters. These stations include a suite of sensors to measure both meteorological and oceanographic variables, such as temperature, solar radiation, wind speed and direction and water salinity. The data are transmitted hourly and automatically analysed for conditions, including high-temperature events, which may cause bleaching or otherwise be stressful to coral reef organisms.

Coral Bleaching Warning is a NOAA web-based warning system that uses a combination of observation stations and satellite-based data to keep track of day-to-day sea surface temperatures in key coral reef regions. Once a region of concern is identified through HotSpots or DHW analyses, or through direct measurements at a CREWS station, a warning is posted on the NOAA website to alert local reef researchers and managers of the threatening conditions.

The coral bleaching phenomenon

Reef corals are animals that house microscopic algae in their tissues in a classic symbiotic relationship. In fact, most corals owe their colours to the presence of these algae and algal pigments. Coral growth rates and health are greatly enhanced by the fact that the solar-powered algae provide food and nutrients to the coral. Under stressful conditions, however, corals (and many other reef organisms that have symbiotic algae, such as sponges and anemones) will expel their algae. Without the algae, the live coral tissue is transparent and the underlying white skeleton shows through, leaving the organism with a "bleached" appearance (fig. 4.4).

Many types of stress can cause corals to bleach, including changes in temperature (both cooling and heating), light intensity and water salinity. Small-scale bleaching events in response to locally intense rainfall (and a sudden drop in salinity), for example, have been observed for many decades. Bleached corals and other organisms can recover from bleaching once conditions return to normal, but corals will die if stressful conditions are prolonged or too severe.

Large-scale bleaching events – those that affect coral reefs over broad geographic regions – were unknown until 1982, when reefs in the eastern Pacific first suffered extensive bleaching and mortality during the major 1982–1983 El Niño event. Since then, considerable bleaching has affected reefs in every region of the tropics and nearly every

Figure 4.4 A bleached coral reef
Source: Dr Ray Berkelmans, provided through ReefBase.

year. Significantly, in the most extensive events, the warmer than normal water temperatures and increased light intensity that occur during El Niño years, when shifting weather patterns lead to warmer and calmer waters, have played a role. During the 1997–1998 El Niño, for example, unusually warm seawater temperatures caused severe mortality in about 15 per cent of reefs worldwide, particularly those in the Indian Ocean. The 2001 bleaching event in Australia's Great Barrier Reef, which was the worst on record for the area, was not related to El Niño, however, but to a general and continuing warming trend in the western Pacific.

Typically, natural selection over long periods favours individuals and species that are best adapted to environmental changes such as temperature increases. Corals possess a variety of mechanisms evolved over millennia that have enabled them to survive environmental fluctuations for millions of years, but has been happening in the oceans over the past few decades is different. Ocean *hot spots* have been causing significant short-term spikes in temperatures that exceed the evolved capacities of many corals to adapt, resulting in higher than natural rates of mortality.

Future coral bleaching events are likely if sea surface temperature increases continue to outpace the speed at which corals can adapt. Although nothing can be done to prevent bleaching when reefs are overstressed, reducing adverse impacts from overfishing, pollution, sedimentation and direct human contact can improve rates of both coral and ecosystem recovery. Furthermore, early warning systems for coral bleaching allow managers to take steps to reduce the impacts of modern stresses on reefs.

Information about NOAA's Coral Health and Monitoring Program (CHAMP) and its Integrated Coral Observing Network (ICON) is available at www.coral.noaa.gov/crews/index.shtml. Tropical ocean coral bleaching indices are available at www.osdpd.noaa.gov/PSB/EPS/CB_indices/coral_bleaching_ indices.html.

Sea-level rise

Susanne Moser

If the earth were perfectly still – no winds, no rotation, no movement of tectonic plates, etc. – water in the oceans would only be affected by gravity pulling it towards the planet's centre. The resulting surface of the oceans, termed the geoid, would be the mean sea level.

The earth is not perfectly still, of course, and neither is the level of the oceans constant. In fact, sea levels fluctuate on time scales of seconds to millennia to millions of years. Depending on the time scale, different natural processes are predominantly responsible for this rise and fall of the oceans.

- *Long-term sea-level changes* (over millions of years) are caused mainly by tectonic processes that lead to the movement of continental and oceanic plates. Such movements change the shape and volume of ocean basins, induce volcanic activity and increase the production of "juvenile" water. These movements can also contribute to large-scale, long-term shifts in the global climate, which in turn continue to affect sea levels.

- *Medium-term sea-level changes* (over hundreds and thousands of years) relate primarily to the loading (during glacial periods) and unloading (during interglacial periods) of vast ice masses from the continents and to the addition or subtraction of water to ocean basins, the resulting water loads weighing down or lightening up these basins. Many of these processes reflect changes in the earth's atmosphere and climate, which can be caused by changes in the constellation of the sun and planets relative to the earth or by shifts in the tilt and precession of the Polar axis. In modern times, humans have also had significant impacts on the earth's land areas, vegetation, freshwater resources and atmosphere that have changed the climate and affected sea levels on this scale.

- *Short-term sea-level changes* (over hours to years) have terrestrial, lunar, solar and planetary origins. Examples include hourly to daily fluctuations reflecting changes in atmospheric pressure, wind or seismic activity (e.g. tsunamis, nor'easters); monthly changes related to the lunar cycle, occasionally magnified by particular sun/moon/earth alignments; seasonal changes such as periodic shifts in ocean circulation (e.g. ENSO) and accompanying wind reversals; and multi-year shifts associated with various planetary alignments.

Short- and medium-term changes are of greatest relevance to humans and their uses of the immediate coastal zone, with shorter-term sea-level changes being most important because they are often accompanied by violent coastal hazards such as hurricanes, tsunamis or seasonal fluctuations of up to a metre or more that frequently impede water-dependent coastal activities such as boating, fishing or shipping. Of less immediate but no less serious concern are quasi-permanent changes that can lead to coastal erosion (fig. 4.5) and permanent inundation of coastal lands, curtailment or permanent loss of coastal land uses and intrusion of salt water into freshwater aquifers (Carter, 1988). The long-term rise in sea level expected with global warming is one example.

Global warming and sea-level rise

Globally, sea levels during the twentieth century rose at a rate of about 1–2 mm per year. With additional climate warming, as projected by the Intergovernmental Panel on Climate Change (IPCC), global sea levels are predicted to rise by from 9 to 88 cm by 2100, a rate roughly two to four times the historical rate. The main contributors to this rise in sea level

Figure 4.5 The rate of coastal erosion is about 100 times that of sea-level rise. Rising water causes beaches to recede and makes structures near them much more vulnerable to storm damage
Source: Jet Propulsion Laboratory (JPL).

are the thermal expansion of seawater as temperatures rise and the melting of inland glaciers and icecaps, and related changes to the storage of water on land.

What rise in sea level will be experienced in any one coastal location as measured at a local tide gauge depends on a number of factors, including:

- the actual global (eustatic) rise in sea level
- regional influences on climate (hence sea level), such as regional warming, the shape and depth of the coastal ocean, etc.
- local vertical movements of the land.

Typical impacts of sea-level rise

Higher sea levels result in a multitude of impacts depending on coastal geology, land forms, the degree of human development and uses of coastal areas, and the degree and type of shoreline protection. The following impacts are typical:

- increased coastal erosion and shoreline retreat rates
- increased storm surges and flooding during coastal storms
- increased saltwater intrusion into estuaries and coastal aquifers
- permanent inundation of previously dry land (see, for example, fig 4.6)
- more frequent or permanent inundation of wetlands, forcing their retreat landward (where landward retreat proves impossible, wetlands will be lost).

Distinctive higher-order impacts include:

- more frequent storm damages to coastal structures and homes, as well as to other structures previously outside the coastal flood zone
- beach losses, especially along structurally protected shorelines, with resulting impacts on coastal properties, roads and other structures as well as on coastal recreation and tourism
- more frequent need for water supply and sewer infrastructure upgrades, relocations and repairs
- decreased shellfish harvests where wetlands are being or have been lost.

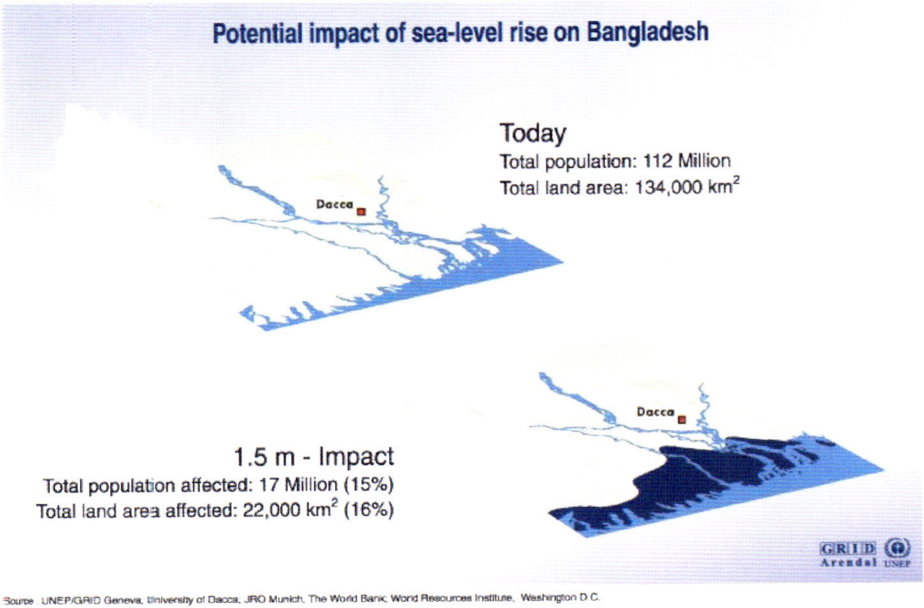

Figure 4.6 Potential impact of sea-level rise on Bangladesh
Source: UNEP/GRID-Arendal Maps and Graphics Library.

What happens on land when global sea level rises? Four scenarios:

Global sea level rising *Land rising at same rate*	No *relative* sea-level change
Global sea level rising *Land rising at a faster rate*	Relative sea-level *falls*
Global sea level rising *Land not rising or falling*	Relative *local* sea-level rises
Global sea level rising *Land falling*	*Faster-than-global* sea-level rises

Integrated Vector Management

Poorly designed irrigation and water systems, inadequate housing, poor waste disposa and water storage, deforestation and loss of biodiversity all may be contributing factors in the spread of the most common vector-borne diseases, including malaria, dengue anc leishmaniasis. Malaria, the most deadly vector-borne disease (fig. 4.7), kills over 1.2 millior people annually, mostly African children under the age of five. Dengue fever, with relatec dengue haemorrhagic fever (DHF), is the world's fastest-growing vector-borne disease.

New strategies for prevention and control of these and other vector-borne diseases emphasize Integrated Vector Management (IVM) as an approach to reinforcing linkages be-

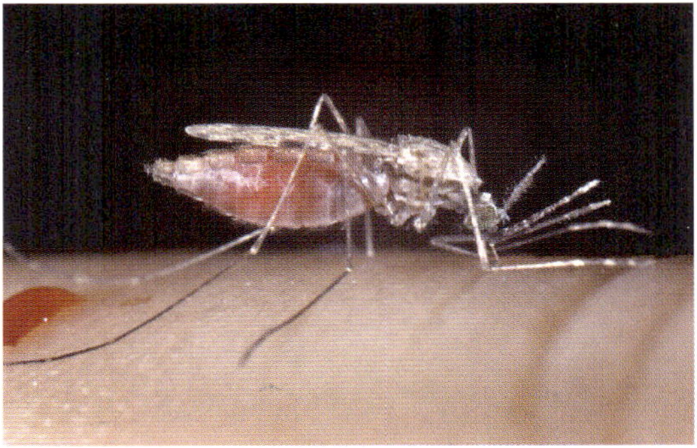

Figure 4.7 A disease vector, the female *Anopheles gambiae* mosquito, feeding
Source: Jim Gathany/CDC (USA).

ween health and environment, optimizing benefits from both. The World Health Organzation (WHO) and the UN Environment Programme (UNEP), for example, have jointly created the Health and Environment Linkages Initiative (HELI). The following overview of VM focused on vector-borne disease is taken from its site.

VM strategies are designed to achieve the greatest disease control benefit in the most cost-effective manner, while minimizing negative impacts on ecosystems, such as depleion of biodiversity, and adverse side-effects on public health from the excessive use of chemicals in vector control. Rather than relying on a single method of vector control, IVM stresses the importance of first understanding the local vector ecology and patterns of disease transmission, and then choosing appropriate vector control tools from the range of options available.

Among these options are *environmental management* strategies that can reduce or eliminate vector breeding grounds altogether through improved design and operation of water resource development projects, as well as use of *biological controls* such as bacterial larvicides and larvivorous fish that target and kill vector larvae without generating the negative ecological impacts of chemicals. When other measures prove ineffective or not cost-effective, however, IVM does make judicious use of *chemical* means of vector control, such as indoor residual sprays, space spraying and chemical larvicides and adulticides that reduce disease transmission by shortening or interrupting the lifespans of vectors.

Regardless of the chosen tools, IVM requires a multisectoral approach to vector-borne disease control. For instance, Health Impact Assessments (HIAs) of new infrastructure development, such as water resources, irrigation systems and agricultural improvements help identify potential impacts on vector-borne diseases upstream of major policy decisions so that effective proactive actions can be taken.

IVM should not be construed as a panacea; however, in many settings the use of IVM strategies has yielded sustainable reductions in disease and transmission rates.

5

Earth hazards

Tsunamis

Ian Kelman and S. H. M. Fakhruddin

The Pacific Tsunami Warning Center (PTWC) in Hawai'i provides *tsunami messages* to most Pacific islands, the Indian Ocean region and American islands in the Caribbean. The West Coast/Alaska Tsunami Warning Center (ATWC) in Alaska provides *tsunami messages* to coastal Alaska, British Columbia, Washington, Oregon and California.

- *Tsunami Information Bulletin or Tsunami Information Statement:* Indicates that an earthquake has occurred and describes its tsunami-generating potential, often stating that no threat exists and thus averting unnecessary evacuations from coastal areas. A bulletin or statement may, if appropriate, be upgraded to a watch or warning.

- *Tsunami Advisory:* The third-highest level of tsunami alert, issued to areas not currently in either warning or watch status when a warning has been issued for another region. Indicates that an area is outside a region under a current warning/ watch.

- *Tsunami Watch:* The second-highest level of tsunami alert, issued based on seismic data without confirming that a destructive tsunami has formed. Provides advanced alert that a destructive tsunami might be forming or already approaching.

Heads Up! Early Warning Systems for Climate-, Water- and Weather-Related Hazards, Michael H. Glantz (ed.), United Nations University Press, 2009, ISBN 978-92-808-1169-8

- *Tsunami Warning:* The highest level of tsunami alert, indicating that actions such as evacuations of low-lying coastal zones in the warned area and relocation of boats and ships to deep water should be taken in response to a tsunami threat. Provides advanced alert that a destructive tsunami has formed and is approaching the warned area.

- *Warning Cancellation:* A final text product indicating the end of a tsunami threat usually issued after ocean data have confirmed that the warned area is no longer threatened.

Table 5.1 Tsunami warning system

Mw less than 6.5 ***(Mw: Moment Magnitude)***	**Earthquake Message Only**
Mw 6.5 to 7.5	**Tsunami Information Bulletin**
Mw 7.6 to 7.8	**Regional Tsunami Warning (1000-km Limit)**
Mw > 7.8	**Expanding Warning / Watch**
Confirmed Teletsunami	**Pacific-Wide Warning**

Source: McCreery (2006).

- *Final Warning Supplement:* A final text product issued following a tsunami episode that may continue to pose a threat. Provides guidance to local officials regarding when they can consider the threat to have passed; only local officials, however, can issue an all-clear for their area.

Indian Ocean tsunami 2004

At 7:58 am local time on 26 December 2004 a huge-magnitude earthquake struck at a depth of 30 km off the northwestern coast of Sumatra, near the small island of Simeulue in Aceh Province, Indonesia. Within an hour tsunamis had killed over 150,000 people along Sumatra's coastline and were racing as swiftly as 805 km/h (500 mph) across the Indian Ocean. Seven hours after the earthquake, tsunamis had slammed into the coasts of more than a dozen other countries, killing an additional 100,000 people.

Several centres that monitor earthquakes at both international and national levels quickly recognized that a big undersea seismic event with significant tsunami potential had struck that morning, but as the destructive waves bore down upon coasts around the Indian Ocean, no formal tsunami warning system was in place to alert the region to the approaching hazard. Immediately following this disastrous event, international, regional and national efforts intensified to establish an end-to-end tsunami warning and mitigation system, and within six months, the Intergovernmental Coordination Group for the Indian Ocean Tsunami Warning and Mitigation System (ICG/IOTWS) was established as the regional body charged with coordinating the design and implementation of an effective and durable system.

Why had there been no warning system in place on 26 December? Despite more than 30 years of effort trying to create an Indian Ocean tsunami warning system, no agency or government had been willing to pay or take responsibility for such a system. Without

Figure 5.1 Tsunami damage in Banda Aceh, Indonesia, in 2004
Source: FAO.

monitoring equipment, trained personnel and local tsunami awareness and plans in the countries bordering the Indian Ocean, viewing the seismic data from afar did not give a clear picture of what was happening, how destructive the tsunami was or how to inform people in danger and encourage them to act appropriately. Unfortunately, the extent of

Figure 5.2 Damage to fishing boats from the December 2004 tsunami
Source: FAO.

the disaster in the Indian Ocean (figs. 5.1 and 5.2) and the continuing threat further to the west was obvious only after most deaths had already occurred. At that point, improvisation was necessary to contact proper authorities and evacuate coastal areas that remained threatened. Indeed, such impromptu actions probably saved scores of lives in eastern Africa.

In fact, numerous informal warning systems at local levels were later identified as having effectively alerted populations of impending crises. In Thailand, for example, one man saw the tsunami hit the coast near him and phoned his brother further along the coast to tell him what was about to happen, while in Sri Lanka a seismologist who happened to be on the beach identified the early signs of the destructive waves and warned everyone to evacuate just before the beach was inundated. On Simeulue (Indonesia), experiences of the devastating 1907 tsunami had been passed along for decades through oral history, so when the earth shook, the population knew to run away from the coastline. In the end, only seven people perished out of over 78,000 islanders even though the waves hit less than 10 minutes after the earthquake.

One lesson of these anecdotes is that individual, local and indigenous knowledge of potential environmental threats and appropriate actions in different circumstances must be integrated into more technical approaches and international coordination plans to produce effective warning systems that are relevant to local populations.

Volcano early warning systems

Ilan Kelman

As part of a national effort to adopt common terminology when describing the state of unrest at American volcanoes, the Alaska Volcano Observatory implemented a new volcanic alert-level system, the Alaska Volcano Warning System, in 2006. The following terms of alert are assigned based on observations, instrumental monitoring data and the known history and potential hazards of each volcano.

- *Normal:* Typical background activity of a volcano in a non-eruptive state; or, after a change from a higher level, when volcanic activity is considered to have ceased.

- *Advisory:* Elevated unrest above known background activity; or, after a change from a higher level, when volcanic activity has decreased significantly, though it continues to be closely monitored for possible increase.

- *Watch:* A volcano is exhibiting heightened or escalating unrest with increased potential for eruptive activity; or a minor eruption is occurring that poses limited hazards.

- *Warning:* An extremely hazardous eruption is occurring or is imminent.

Alert-level announcements will always contain additional explanations of volcanic activity and the expected hazards.

Volcanoes produce a variety of hazards, impacting people, ecosystems, structures, weather, climate and even aircraft at cruising altitude. Volcanic hazards include solids such as rocks thrown for kilometres, liquids such as lava (molten rock) and poisonous gas emissions like sulphur dioxide. Mixtures of these hazards also occur – for example, the solid-liquid mix of lahars (mudflows) and the gas-dust mix of pyroclastic flows – as do energy hazards such as explosion shockwaves and volcano-created lightning. Ground movement can prove problematic, too, through earthquakes, subsidence, inflation and surface deformations.

Because of the variety of volcanic hazards and the strong differences between different volcanoes and volcanic events, developing a single, effective early warning system or early warning template for volcanoes is difficult. In reality, volcanic warning systems have tended to be most effective when adapted to specific volcanic locations, hazards or potential incidents.

Figure 5.3 Alaska's Mt. Redoubt became active in March 2009, spewing ash and causing small, shallow volcanic earthquakes
Source: Alaska Volcano Observatory.

The US Geological Survey (USGS) is the government agency most involved in volcano research and monitoring, both domestically and internationally. In addition to general volcano warnings, the USGS has developed several warning systems for specific volcanic hazards (see http://volcanoes.usgs.gov for the Volcanic Hazards Program). For example, automated warning systems for lahars (mudflows) have been established at Mount Rainier in Washington State and Mount Redoubt in Alaska.

In an important test this system detected, warned of and tracked lahars moving along the route they were expected to follow down Alaska's Mount Redoubt in 1990. Because of the warning system, the speed of the lahars could be calculated, although their size proved difficult to determine accurately. Even so, if people had been in danger they would have had enough warning time to evacuate, as was the case with a devastating lahar in Armero, Colombia, on 13 November 1985. Although over 22,000 were killed, numerous survivors reported that a warning of only a few minutes had provided them with enough time to scramble to higher ground. They lost everything but their lives.

Development of the Mount Pinatubo Warning System

In April 1991, Mount Pinatubo in the Philippines began rumbling. Drawing upon long-standing relationships with the USGS, Philippine volcanologists requested help and, in the weeks prior to the climactic eruptions in mid-June (fig. 5.4), worked together with US scientists to develop a six-level alert system that included procedures for downgrading alerts.

Pinatubo ended up being one of the largest eruptions of the twentieth century, cooling the global climate for some years afterwards because of the materials it injected into the atmosphere. Careful monitoring, however, and the warning system that had been established only weeks prior to the eruption enabled a staged evacuation that effectively moved hundreds of thousands of people away from the principal danger zone.

Figure 5.4 The first major eruption of Pinatubo in 1991 occurred on 12 June. The column in the eruption phase rose about 20 km. For more details on Pinatubo see http://vulcan.wr.usgs.gov/Vdap/Responses/Pinatubo91/pina_yrbk-1991.html
Source: R. S. Culbreth, US Air Force.

Even so, the number of deaths indirectly attributed to Pinatubo's eruption is regrettable. Several hundred people were killed, for instance, when a typhoon's rainfall collapsed ash-laden roofs where people had sheltered from the eruptions. Fatalities in poorly managed evacuation camps continued to mount for weeks after the main eruptions, and for more than a decade volcanic mudflows claimed lives in the surrounding areas. Without the comprehensive warning system implemented in the weeks prior to the eruption, however, experts surmise that the death toll may have been in the tens of thousands.

The Pinatubo warning system has now been updated and adapted for other locations. Taal Volcano, also in the Philippines, for example, uses a four-level, more detailed variant of the original Pinatubo EWS that considers new data sources, like the temperature and chemical composition of the lake that fills the volcanic crater, in determining the status level of the volcano. The alert system includes "required actions" for each level, which informs people exactly what they should do in any volcanic circumstance.

Earthquake early warning

Ilan Kelman

Early warning systems for earthquakes are realizable because earthquake waves, which consist of two sets, the primary and the secondary, take time to travel. Primary waves travel fast and arrive at an affected area first, but usually do not cause major damage. Early warning detectors can sense these primary waves and immediately transmit a warning, which enables action before the slower, generally much more damaging, secondary waves hit.

The effectiveness of an earthquake warning depends on the proximity of the quake's epicentre to the affected area. In some cases, the time period between a location's warning and the start of major tremors (i.e. the arrival of the secondary waves) is over a minute, while in other cases, where the epicentre is nearby – or in places where a faultline runs underneath a city – no warning time is possible.

Japan, Mexico and Taiwan, for example, all have early warning systems for earthquakes which, if the epicentre of an event is far enough away, can provide enough warning for people to take shelter, for trains to brake, for traffic lights to turn red to halt traffic and for elevators to open automatically at the nearest floor. Other possible responses include automatic cut-offs for gas and electricity.

Unfortunately, many people might develop a false sense of security when warning systems such as these exist, assuming that they will have 30–60 seconds to act after a warning is broadcast, when in reality they may have much less time than that. Furthermore, as critics of such systems suggest, focusing only on short warning times – an important but, in truth, rather trifling enterprise – can actually distract planners from long-term essentials like designing safe infrastructure and developing community capacity to cope with earthquake hazards.

These critics argue that the best warning for earthquakes is prudent city planning and preparation. Infrastructure must be constructed to withstand major tremors, and people must be trained in correct earthquake behaviour, such as the mnemonic "drop, cover, hold", which prompts people to take quick action by dropping to the floor, taking cover under a sturdy object and holding on during the quake. Such sensible planning would increase the likelihood that even a few seconds of warning would mean greater survival in earthquakes, where people are often injured or killed by failing structures or falling debris as they attempt to exit a building.

With several earthquake early warning systems operational in the world today (fig. 5.5), data will soon be available to determine accurately the effectiveness of such systems after an earthquake has occurred. Only with these data will planners better understand whether a few seconds does indeed save lives, or whether only knowing that a specific fault will rupture at some future date is enough to take appropriate long-term preventive actions to plan for such an inevitability.

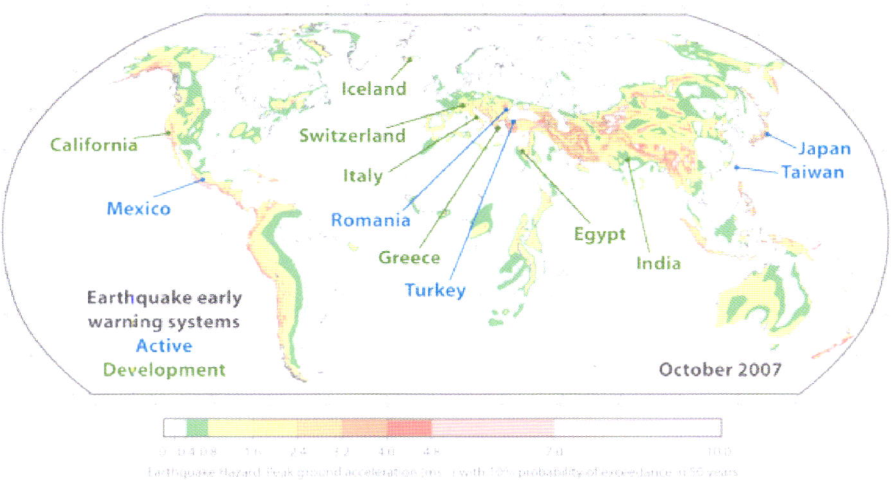

Figure 5.5 Global earthquake hazard map showing regions operating active earthquake EWSs (in blue). The regions developing EWSs are in green
Source: Richard M. Allen, Seismology Lab, UC Berkeley.

Space weather

Thomas Bogdan

The NOAA Space Environment Center (SEC) is the official source for space weather forecasts, alerts and warnings in the United States. As a national critical system, it operates 24 hours a day, 365 days a year without interruption. The SEC products and services are distributed in real time to government agencies, private sector customers and partners around the globe.

Like hurricanes, space weather is classified by a category index ranging from 1 (minor) to 5 (extreme), but for three distinct forms of space weather events: *geomagnetic storms* (G1–G5), *solar radiation storms* (S1–S5) and *radio blackouts* (R1–R5).

- *Geomagnetic storms* describe conditions when the earth's magnetic field fluctuates rapidly in both space and time, usually in concert with enhanced auroral activity and energetic particle precipitation over the Polar Regions. These fluctuating fields drive spurious yet destructive electrical currents along power lines and down oil and gas pipelines. Energetic particles from these storms also produce surface charging on satellites which can lead to electrical discharges that can damage or destroy delicate electronics and sophisticated communications systems.

- *Solar radiation storms* occur when the earth's atmosphere is bombarded with high fluxes of relativistic ions, protons and electrons. Fortunately, our atmosphere stops nearly all of these particles before they reach the earth's surface, but dangerous radiation levels are possible for high-flying aircraft and astronauts operating above the bulk of the atmosphere. Additional impacts associated with radiation storms include disruptions in communication, satellite surface charging and errors in GPS operations.

- *Radio blackouts* occur when the earth's sunlit atmosphere is bombarded with high fluxes of energetic photons (x-rays and extreme ultraviolet radiation). These photons ionize the upper portions of the neutral atmosphere and rapidly enhance the extent of the ionosphere, which can result in the total loss of radio communication capabilities over the dayside hemisphere. Systematic GPS errors

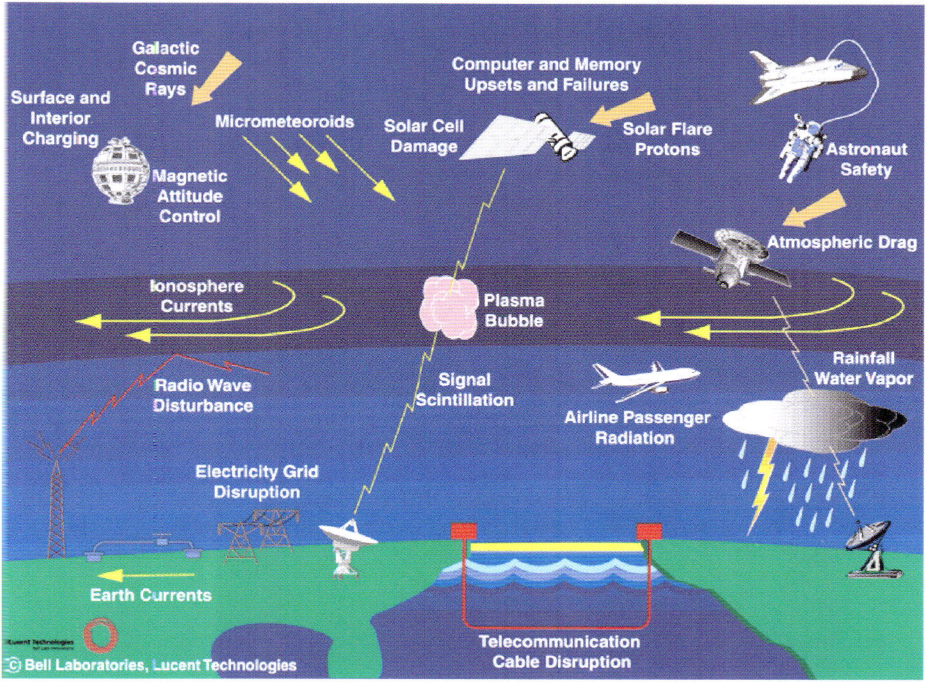

Figure 5.6 Space weather hazards
Source: Lou Lanzerotti, Bell Laboratories.

on the order of 50 m can be induced, and in some extreme cases GPS satellite locks can be lost entirely.

Space weather affects satellite operations, radio communications, numerous diverse enterprises that depend upon precision GPS capabilities, air-traffic control and routing space exploration and tourism, electrical power grids, homeland security and battlespace situational awareness (fig. 5.6). Space weather is also responsible for the exquisite natural phenomena associated with the aurora borealis (the Northern Lights, fig. 5.7) and the aurora australis (the Southern Lights).

Essentially, all space weather originates at the sun, our nearest star. The underlying cause of space weather is the complicated dynamical evolution of the solar magnetic field that is driven by a combination of the sun's differential rotation and its turbulent convection. For reasons we do not fully understand, this combination imprints an approximately 11-year modulation on the frequency of space weather storms that is also reflected in the familiar waxing and waning of the number of *sunspots* visible on the solar disk. Sunspots are merely harbingers of the levels of solar magnetic activity, however and do not directly contribute to solar storms. Rather, solar storms can be traced back to *solar flares* (which may produce earthward-bound relativistic ions, protons and electrons in addition to x-rays and extreme ultraviolet radiation), *coronal mass ejections* (responsible for geomagnetic storms and relativistic particles) and *high-speed solar wind streams* (particles and geomagnetic storms). The solar magnetic field is the energy source and catalyst for all three of these storm-producing agents.

Coronal mass ejections and high-speed streams require from one to three days to propagate the distance from the sun to the earth. Consequently, accurate forecasts are routine for these same time scales when satellite and ground-based observations detect the signatures of genesis of these storms at the sun. For solar flares, on the other hand, because photons travel from the sun to the earth in approximately eight minutes, when

Figure 5.7 The Northern Lights reach across a forest lake in the upper-Midwestern United States
Source: NASA.

a flare is observed on earth the associated radio blackout is already in progress. In essence, we must learn how better to predict when and where a flare will occur if we want better forecasting of radio blackouts.

US ultraviolet forecasts

The US National Weather Service daily forecasts of UV (ultraviolet) levels use the following scale (fig. 5.8).

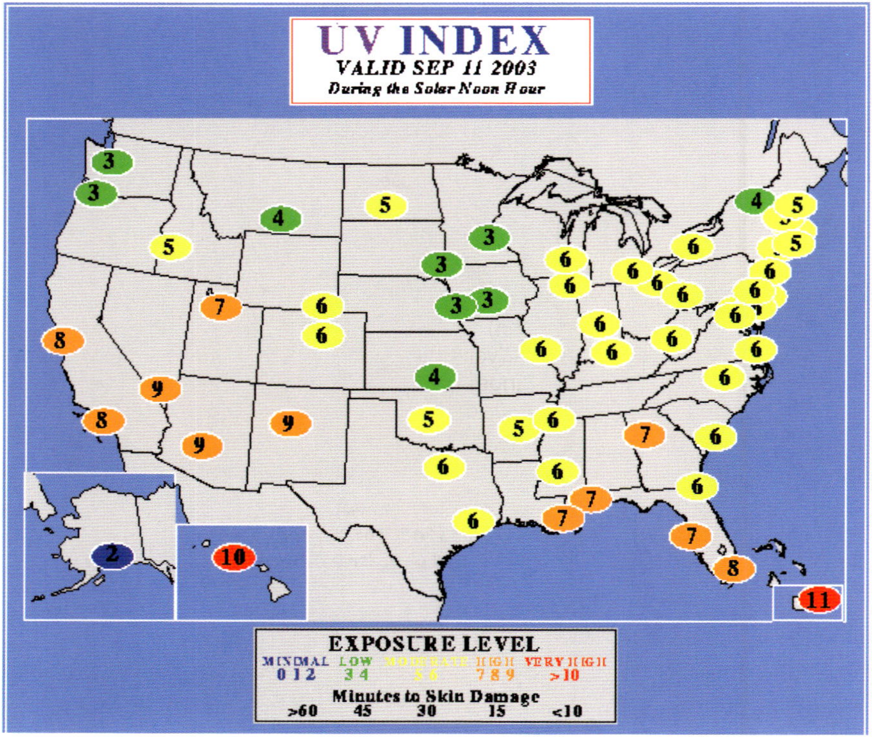

Figure 5.8 UV forecasting in the United States
Source: NOAA.

Table 5.2 UV exposure and preventive action

Level	Time to burn	Actions to take at noon
Minimal	60 minutes	Apply SPF sunscreen
Low	45 minutes	Apply SPF, wear a hat
Moderate	30 minutes	Apply SPF 15, wear a hat
High	15–24 minutes	SPF 15–30, hat, sunglasses. Limit midday exposure
Very high	10 minutes	VSPF 30, hat, sunglasses, protective clothing. Avoid midday exposure

Number	*Exposure level*
0, 1, 2	Minimal (blue)
3, 4	Low (green)
5, 6	Moderate (yellow)
7, 8, 9	High (orange)
10 or higher	Very high (red)

Table 5.2 shows what these levels imply. "Time to burn" and "Actions" apply to people with a type II fair skin that sometimes tans and usually burns. People with lighter skin need to be more cautious; people with darker skin may be able to tolerate more exposure, though even dark skin can burn.

6

Concluding thoughts

Lessons learned about "lessons learned"

International conferences on EWSs are attempts by agencies and governments to keep abreast of new developments in research and operations. Such international meetings also provide a time for retrospection, as case studies of hazard response successes and failures are scrutinized at such conferences. These cases add to the substantial knowledge base that exists about hazards and disasters, and about early warning systems, both formal and informal. They are also important because, even if expected hazards fail to occur, the process of warning itself has unique societal impacts that must be considered.

What just about every one of these hazard- or disaster-related case studies ends with is an assessment of "lessons learned" (I have done this as well – Glantz, 2001). I have now come to believe that the phrase "lessons learned" has become a part of the problem in addressing issues related to early warning and coping with future hazards.

A preliminary review of "follow-ups" (retrospectives) on case-specific lessons learned exposes the reality that many of these often costly lessons were not, in fact, really "learned" in the true sense of the term. A multi-decade review examining previous similar disasters either in the same location or elsewhere would most likely uncover similar, previously identified and reported lessons learned. Again and again, these same lessons are merely identified.

Heads Up! Early Warning Systems for Climate-, Water- and Weather-Related Hazards, Michael H. Glantz (ed.), United Nations University Press, 2009, ISBN 978-92-808-1169-8

While some lessons are actually identified and applied – truly "learned" – as a result of these assessments, many responses to the lessons learned from particular cases go unfunded and are forgotten, only to be rediscovered again after another major disaster when a new series of assessments identifies and reports the same lessons. And so the cycle continues.

The phrase "lessons learned" suggests to the public that someone, some unknown person or agency or government department, is taking care of a concern. Unfortunately, however, no one may have been delegated to take on *that* responsibility, and as time passes interest in that specific disaster and its victims wanes, overshadowed by other pressing issues of the day and by the impacts of more recent disasters.

Two significant questions arise. Once lessons have been identified and publicly broadcast, who is expected to take action to make sure those lessons *are* truly learned? And are those who initially identified the lessons in a position to influence those who are in a position to implement the changes prescribed by the lessons?

Perhaps an assessment should be commissioned to identify and expose the obvious as well as hidden reasons why hazard- and disaster-related lessons are so regularly identified, only to end up gathering dust on a bookshelf, never to be applied.

Foreseeability of hazards

Long before remote sensing and other electronic technologies were developed to monitor for early warning, environmental signs as well as historical remembrances and intuitions that were passed orally from one generation to the next were used as indicators of environmental change.

"Canary in the Coal Mine"

"The canary is particularly sensitive to toxic gases such as carbon monoxide which is colourless, odourless and tasteless.

This gas could easily form underground during a mine fire or after an explosion.

Following a mine fire or explosion, mine rescuers would descend into the mine, carrying a canary in a small wooden or metal cage.

Any sign of distress from the canary was a clear signal the conditions underground were unsafe and miners should be evacuated from the pit and the mineshafts made safer."

On This Day, 30 December, BBC

Canaries were used in English mines until the mid-1980s.

Pratt (2002), for example, suggests that pastoralists in Africa develop warnings based on several considerations: they have an understanding of the probability of future rain by variances in wind, humidity and temperature from expected conditions; they know how to interpret the behaviour of animals and plants, which serve as valuable indicators for subtle fluctuations in temperature and humidity; and they have observed historical trends that allow for reasonable predictions of future weather patterns. (Notably, Pratt goes on to say that, because of global warming, past trends may soon no longer be useful for projecting future conditions.)

In general, research into informal warning systems shows that societies have always learned from their environments about the *foreseeable* hazards they believed they might have to face. These were not, of course, perfect systems (the high-tech systems we rely on today also lack perfection), but they were responses to the natural hazards local communities encountered seasonally, annually or just occasionally.

The operative term here is "foreseeable". The concept of "foreseeability" has been used for over a century as a legal test for liability. It is "a concept used in various areas of the law to limit the liability of a party to the consequences of his/her actions that were within the scope of a foreseeable risk". As Gifis (1991) asserts, "The foreseeability element of proximate cause is established by proof that the actor or person of reasonable intelligence and prudence should reasonably have anticipated danger to others created by his or her negligent act".

Important to keep in mind is that inaction is considered a form of action – the negligence referred to here being that of an authority which failed to foresee adverse impacts and act to mitigate or prevent them. As Gifis (ibid.) notes: "Foreseeability encompasses not only that which the defendant foresaw, but that which the defendant ought to have foreseen."

"Foreseeability" differs from "forecastability" or "predictability" because it does not depend on any quantitative assessment of the probability of occurrence. Instead, foreseeability suggests that a reasonable person can conclude that certain impacts from certain types of hazards will have knowable adverse consequences.

For identifying the possible impacts of climate-, water- and weather-related hazards about which both governments and citizens at local to national levels are concerned, the concept of foreseeability can be extremely useful. Also useful is the foreseeability of the likely impacts of demographic trends (population increases and migrations, for instance) on the levels of risk populations may face from local and regional hazards.

As of today, researchers have collected hundreds (probably even thousands) of examples of what happens when a hurricane, severe storm, drought, flood, bushfire or disease outbreak occurs in different parts of the world at different times of the year in different types of societies. A lot of information also exists about both the strengths and the weaknesses of early warning systems for these hazards. We have even calculated with a good measure of accuracy both which and how societies are likely to be at risk to these climate-, water- and weather-related hazards. In addition, we have a reasonable idea about the populations within societies that are most likely to be affected by these hazards' various impacts. Furthermore, a range of coping strategies and tactics to contend with such hazards has been developed, some having been successful and some having failed.

The concept of *foreseeability* should therefore be regarded as an early warning opportunity because of its capacity to attend to and qualify the contexts of the individuals who are potentially in harm's way, and of those responsible for protecting them – much as *probability*, the mathematically based determination of risk, can quantify the impacts of those hazards. For early warning systems, foreseeability and probability should both be considered in determining levels of threat for a given population.

References

All website addresses current as of November 2008.

Carter, R. W. G. (1988) *Coastal Environments: An Introduction to the Physical, Ecological and Cultural Systems of Coastlines*, London: Academic Press, pp. 245–279.

CPC (2006) "Expert Assessments of the Climate Prediction Center, National Weather Service, NOAA", available at www.cpc.noaa.gov/products/assessments/.

Deutsche Presse Agentur (2009) "Myanmar Needs 691 Million Dollars for Post-Cyclone Recovery", Reliefweb, available at www.reliefweb.int/rw/rwb.nsf/db900SID/ASAZ-7P4ET4?OpenDocument.

EEA (2002) *Late Lessons from Early Warnings: The Precautionary Principle 1896–2000*, European Environment Agency, Environmental Issue Report No. 22, available at http://reports.eea.europa.eu/environmental_issue_report_2001_22/en.

FAO (2000) *Report of the 26th FAO Regional Conference for Latin America and the Caribbean*, held in Mérida, Mexico, 10–14 April, FAO Doc. LARC/00/4, pp. 7–8.

Gifis, S. (1991) *Law Dictionary*, 3rd edn, New York: Barron's Educational Series, p. 195.

Glantz, M. H., ed. (2001) *Once Burned, Twice Shy? Lessons Learned from the 1997–98 El Niño*, Tokyo: United Nations University Press.

——— (2005) "Hurricane Katrina Rekindles Thoughts about Fallacies of a So-called 'Natural' Disaster", *Sustainability: Science, Practice & Policy* 1(2), available at http://ejournal.nbii.org/current.html.

Gleick, P. H. (1996) "Water Resources", in S. H. Schneider, ed., *Encyclopedia of Climate and Weather*, New York: Oxford University Press, vol. 2, pp. 817–823.

Gruntfest, E. and A. Ripps (2000) "Flash Floods: Reducing Loss Susceptibility in Flash Floods", in D. Parker, ed., *Floods*, Basingstoke: Routledge, pp. 377–390.

International Strategy for Disaster Reduction (2008) *Indigenous Knowledge for Disaster Risk Reduction: Good Practices and Lessons Learned from Experiences in the Asia-Pacific Region*, Bangkok: United Nations, available at www.unisdr.org/eng/about_isdr/isdr-publications/19-Indigenous_Knowledge-DRR/Indigenous_Knowledge-DRR.pdf.

Klinenberg, E. (2002) *Heat Wave: A Social Autopsy of Disaster in Chicago*, Chicago, IL University of Chicago Press.

Larsen, J. (2006) "Setting the Record Straight: More than 52,000 Europeans Died from Heat in Summer 2003", Washington, DC: Earth Policy Institute, available at www.earth-policy.org/Updates/2006/Update56.htm.

Liu, J., H. Xiao, F. Lei, Q. Zhu, K. Qin, X.-W. Zhang, X.-L. Zhang, D. Zhao, G. Wang, Y Feng, J. Ma, W. Liu, J. Wang and G. F. Gao (2005) "Highly Pathogenic H5N1 Influenza Virus Infection in Migratory Birds", *Science* 309, p. 1206.

MARA (2004) "Project to Provide an Atlas of Malaria for Africa", Mapping Malaria Risk in Africa, available at www.mara.org.za/.

McCreery, Charles S. (2006) "Tsunami System Design and Plan", paper presented a TARNS Workshop, 24–27 May, Hua Hin, Thailand.

National Agricultural Monitoring System (2005) "Homepage", available at www.nams gov.au.

National Meteorological Agency (2007) "Climatological Services Team", *Climate Information for the Health Sector* 1(9), available at www.ethiomet.gov.et/pdf_files/HealthbulletinSept2007.pdf.

Pegg, J. R. (2004) "The Arctic: Earth's Early Warning System", *AlterNet*, 19 September available at www.alternet.org/story/19930/.

Pielke, R. A. Jr. and R. A. Pielke Sr. (1997) *Hurricanes: Their Nature and Impacts on Society*, New York: John Wiley & Sons.

——— (2003) "Hurricane as an Extreme Meteorological Event", in *Handbook of Weather Climate, and Water: Atmospheric Chemistry, Hydrology, and Societal Impacts*, New York: John Wiley & Sons.

Pratt, C. (2002) *Traditional Early Warning Systems & Coping Strategies for Drought Among Pastoralist Communities*, Somerville, MA: Tufts University, Feinstein International Famine Center, Fletcher School.

Storm Prediction Center (2006) NOAA's National Weather Service real-time service reporting US thunderstorms, available at www.spc.noaa.gov.

Tannehill, I. R. (1947) *Drought and Its Causes and Effects*, Princeton, NJ: Princeton University Press.

US Geological Survey (2006) "The Earth's Water Cycle", available at http://ga.water.usgs.gov/edu/watercyclesummary.html.

Weather Network (2007) "International Weather Forecasts", issued down to a 10 square kilometre grid, in English and French, available at www.theweathernetwork.com.

Other resources for early warning systems

NOAA National Weather Service Forecast Office, Chicago, IL: "Weather Terms and Warning/Advisory Criteria", available at www.crh.noaa.gov/lot/severe/wxterms.php#gwt.

National Hurricane Center, Tropical Prediction Center: "Glossary of NHC/TPC Terms", available at www.nhc.noaa.gov/aboutgloss.shtml.

National Hurricane Center, Tropical Prediction Center: "The Saffir-Simpson Hurricane Scale", available at www.nhc.noaa.gov/aboutsshs.shtml.

NOAA National Weather Service Forecast Office, Peachtree City, GA: "Definition of Flash Flood Statement", available at www.srh.noaa.gov/ffc/html/whatisffs.shtml.

Novy, Chris Hayes, Roger Edwards, David Imy and Stephen Goss: "Heavy Rain MCDs", Southern Illinois University and Storm Prediction Center, available at www.spc.noaa.gov/misc/about.html#Heavy%20Rain%20MCD.

NOAA National Weather Service Climate Prediction Center: "Explanation of Palmer Drought Severity Indices", available at www.cpc.ncep.noaa.gov/products/analysis_monitoring/cdus/palmer_drought/wpdanote.shtml.

NOAA National Weather Service Storm Prediction Center: "Fujita Tornado Damage Scale", available at www.spc.noaa.gov/faq/tornado/f-scale.html.

Government of Malaysia: "Fire Danger Rating System", available at www.kjc.gov.my/english/service/climate/fdrs1.htm.

Canadian Forest Service, Natural Resources Canada: "Canadian Forest Fire Weather Index (FWI) System", available at http://cwfis.cfs.nrcan.gc.ca/en/background/bi_FWI_summary_e.php.

AirNow: "Air Quality Basics and Products", available at http://airnow.gov/.

Colorado Department of Public Health and Environment: "Denver Air Pollution Forecast and Information", available at http://apcd.state.co.us/psi/advisory.phtml.

Ministry of the Environment, Government of Japan: "Dust and Sandstorms (DSS)" available at www.env.go.jp/en/earth/dss/.

NOAA National Weather Service Space Environment Center: "Current Space Weather Conditions and Products", available at www.sec.noaa.gov/index.html and www.sec.noaa.gov/forecast_verification/index.html.

NOAA National Weather Service Pacific Tsunami Warning Center: "Product Definitions" available at www.prh.noaa.gov/ptwc/definitions.htm.

Intergovernmental Oceanographic Commission: "IOC Member States Site", available at http://ioc.unesco.org/iocms/files.php?action=viewfile&fid=190&fcat_id=25.

US Geological Survey: "Volcano Hazards Program: Volcano Warning Scheme for the U.S.", available at http://volcanoes.usgs.gov/Products/Warn/WarnSchemes.html.

US Environmental Protection Agency: "SunWise Program: UV Index", available at www.epa.gov/sunwise/uvindex.html.

Army Study Guide: "Leadersbook Info, Air Defense Warning System", available at www.armystudyguide.com/content/Leadersbook_information/Combat_Leaders_Guide/air-defense-air-defense-w-2.shtml.

Australian Government Bureau of Meteorology: "Bureau of Meteorology Warning Services – Tropical Cyclone, Fire Weather, Severe Weather, Flood Weather, Marine Weather, Tsunami, Road Weather, Agricultural, Haze, Walking", available at www.bom.gov.au/catalogue/warnings/.

Australian Government Bureau of Meteorology: "Rainfall Deficiency Monitoring System", available at www.bom.gov.au/climate/drought/drought.shtml.

UK MetOffice: "Key to Warnings – Gales, Storms, Snow, Fog, Heat, Roads", available at www.metoffice.gov.uk/weather/uk/guide/key_warnings.html.

Australian Department of Agriculture, Fisheries and Forestry: "National Drought Assistance Measures", available at www.daff.gov.au/droughtassist.